T0295203

Negative Capacitance Field Effect Transistors

This book aims to provide information in the ever-growing field of low-power electronic devices and their applications in portable devices, wireless communication, sensor, and circuit domains. *Negative Capacitance Field Effect Transistors: Physics, Design, Modeling and Applications* discusses low-power semiconductor technology and addresses state-of-the-art techniques such as negative capacitance field effect transistors and tunnel field effect transistors. The book is split into three parts.

The first part discusses the foundations of low-power electronics, including the challenges and demands and concepts such as subthreshold swing. The second part discusses the basic operations of negative capacitance field effect transistors (NC-FETs) and tunnel field effect transistors (TFETs). The third part covers industrial applications including cryogenics and biosensors with NC-FET.

This book is designed to be a one-stop guide for students and academic researchers, to understand recent trends in the IT industry and semiconductor industry. It will also be of interest to researchers in the field of nanodevices such as NC-FET, FinFET, tunnel FET, and device–circuit codesign.

Materials, Devices, and Circuits: Design and Reliability
Series Editor: Shubham Tayal, K. K. Paliwal, Amit Kumar Jainy

Tunneling Field Effect Transistors: Design, Modeling, and Applications
Edited by T.S. Arun Samuel, Young Suh Song, Shubham Tayal, P. Vimala and Shiromani Balmukund Rahi

For more information about this series, please visit:
www.routledge.com/Materials-Devices-and-Circuits/book-series/MDCDR

Negative Capacitance Field Effect Transistors

Physics, Design, Modeling and Applications

Edited by
Young Suh Song, Shubham Tayal,
Shiromani Balmukund Rahi and
Abhishek Kumar Upadhyay

CRC Press
Taylor & Francis Group
Boca Raton London New York

CRC Press is an imprint of the
Taylor & Francis Group, an **informa** business

Designed cover image: Shutterstock

First edition published 2024
by CRC Press
6000 Broken Sound Parkway NW, Suite 300, Boca Raton, FL 33487-2742

and by CRC Press
4 Park Square, Milton Park, Abingdon, Oxon, OX14 4RN

CRC Press is an imprint of Taylor & Francis Group, LLC

© 2024 selection and editorial matter, Young Suh Song, Shubham Tayal, Shiromani Balmukund Rahi, and Abhishek Kumar Upadhyay; individual chapters, the contributors

ISBN: 978-1-032-44531-1 (hbk)
ISBN: 978-1-032-44684-4 (pbk)
ISBN: 978-1-003-37339-1 (ebk)

DOI: 10.1201/9781003373391

Typeset in Times
by Newgen Publishing UK

Contents

Preface

Almost on a regular basis, the semiconductor industry has gone through technological innovation, thereby accomplishing significant developmental milestones. The development of metal-oxide-semiconductor field effect transistor (MOSFET) and the advent of the three-dimensional (3D) MOSFET structure have led to the recent state-of-the-art semiconductor production. However, simple structural development has been faced with unavoidable limitations, especially since the size of transistors cannot be easily reduced and the structural development has been almost saturated after the advent of gate-all-around (GAA) MOSFET structure. Fortunately, the modern semiconductor industry has successfully overcome this limitation by material innovation, namely, negative capacitance (NC) technology. With the help of the wonderful combination of nanomaterials, it is entirely possible to boost the transistor's performance and circuit performance at the same time. Moreover, the application of NC technology could boost the performance of the transistor, without reducing its size.

This book, *Negative Capacitance Field Effect Transistors: Physics, Design, Modeling and Applications*, presents a self-contained and cutting-edge technology and sine qua non of ideas and illustrations that will help readers to easily learn about nanomaterials and their applications. This book can also serve as a reference for material engineers and electronic engineers and experts in the semiconductor industry.

The book also carefully considers the basic concepts and physics which are critical to understanding the basic operation principles of modern semiconductors. With the recently advanced quantum mechanics, the overall flow of semiconductor design and development can be easily and intuitively understood. The editors sincerely hope that these up-to-date theories and technological descriptions will enable readers to gain a thorough understanding of modern semiconductors. This book also considers the circuit perspective, so that a more hands-on understanding can be gained.

We, the editors, would like to express our sincere gratitude for all the assistance provided by various research teams and authors in writing and discussing the content of this book and their invaluable data inputs. Also, we are grateful for the time and efforts that all our colleagues and co-workers have invested in the development and design of this book.

Seoul, Republic of Korea
2023

Young Suh Song
Shubham Tayal
Shiromani Balmukund Rahi
Abhishek Kumar Upadhyay

1 Recent Challenges in the IT and Semiconductor Industry
From Von Neumann Architecture to the Future

Young Suh Song[1], Shiromani Balmukund Rahi[2],
Navjeet Bagga[3], Sunil Rathore[3],
Rajeewa Kumar Jaisawal[3], P. Vimala[4],
Neha Paras[5], K. Srinivasa Rao[6]

[1]Department of Computer Science, Korea Military Academy, Seoul, South Korea
[2]Department of Electrical Engineering, Indian Institute of Technology, Kanpur, India
[3]VLSI Design and Nano-Scale Computational Lab, PDPM Indian Institute of Information Technology Design and Manufacturing Jabalpur, Jabalpur, India
[4]Electronics and Communication Engineering, Dayananda Sagar College of Engineering, Bengaluru, India
[5]Department of Electronics and Communication Engineering, NIT Delhi, India
[6]KL University, Green Fields, Vaddeswaram, Andhra Pradesh, India

1.1 INTRODUCTION

We are now living in an unprecedented society, where fifth-generation (5G) technology enables people to communicate with each other at any time, no matter where they are [1]. In addition, the emerging big data market has also enabled emerging industries (e.g., Facebook, Instagram, YouTube, TikTok), thereby creating numerous new jobs [2]. In this regard, the importance of recent semiconductor technology will be crucial, especially since most emerging industries are closely related with semiconductors. Without the recent advances in the semiconductor industry, the above-mentioned industries might not have been able to be created.

According to recent surveys, the size of semiconductor industries has been steadily increased, roughly 20–30% per year [3]. In addition, according to the International

DOI: 10.1201/9781003373391-1

1

Roadmap for Devices and Systems (IRDS) organization, this trend will continue until 2030, with the help of the consistent development of transistors [4]. Therefore, what does the term 'semiconductor' specifically mean? To find the answer, we first need to briefly look into the structure of a computer.

1.2   BASIC ARCHITECTURE OF A COMPUTER

Figure 1.1 shows the image of computer users. Normally, people use a mouse and keyboard while using a computer. If users push any buttons on the keyboard or any buttons of the mouse, then the result will be shown on the monitor. Sometimes, the speaker will be turned on so that some music or soundtracks can be played through a speaker.

In this example, we can easily understand the basic structure of a computer. In this example, 'keyboard' and 'mouse' are examples of input devices, and 'monitor' and 'speaker' are examples of output devices.

This explanation is illustrated in Figure 1.2. Input devices and output devices are somewhat familiar concepts and easy for us to understand. We can see the input devices and output devices whenever we use a computer. Then, what is a 'central processing unit (CPU)' as shown in Figure 1.2? When a computer user uses a computer, the CPU will undergo some operations. For example, when the user clicks the 'Internet button', then the CPU will carry out some operations so that the Internet page can be directly viewed on the monitor. When the computer user clicks the 'My favorite music.mp3' button twice, then the CPU will find that file and play the file right away, so that the output device (speaker) can play the music file.

Let's carefully look at Figure 1.2 again. The CPU consists of a control unit, arithmetic logic unit (ALU), and cache memory. Here, the term 'ALU' means the device which carrier out the basic operation of calculation. For example, ALU does an addition operation (e.g., 'binary adder': 101 + 010 = 111) and deduction operation.

ALU also does logic operations. As shown in Figure 1.3, there are four main logic operations in computer: NOT, AND, OR, and XOR. First, the NOT operator simply reverses the binary value; from '0' to '1', and from '1' to '0'. Second, the AND operator makes an output '1' only if two inputs are both '1' (Figure 1.3). If one of the two inputs is '0' or both of the inputs are '0', the output will be '0'. The OR operator has another function. The OR operator gives '1' output, if there is any '1' among the inputs. For example, when input '1' and '0' flow into the OR operator, the output will be '1'. On the other hand, when only input '0' and '0' flow into the OR operator, the output will be '0'.

So far, we have learned the basic concepts with theoretical approaches. Now, let's move on to a 'hands-on' explanation. Figure 1.4 illustrates the basic structure of computers. Most readers will be familiar with this figure. If you are not familiar with this image, don't worry. When you remove the outer cover of your computer, you can easily identify a similar structure to that in Figure 1.4.

In Figure 1.4, there are a CPU (circle), main memory (rectangle), and memory slot (diamond). The memory slot (diamond) is the place where the main memory device needs to be put in. When the main memory device is successfully put into the memory slot (rectangle), then the computer will be able to save information.

FIGURE 1.1   A person using a computer.

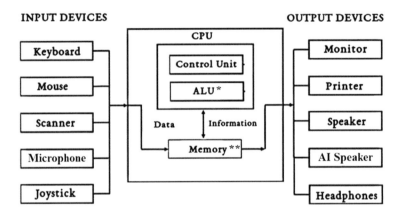

> *ALU stands for Arithmetic Logic Unit, which plays important roles as adder, multiplier, logic operator*
> *** Memory consists of main memory (RAM) and auxiliary memory (Hard disk, CD, DVD, SSD)*

FIGURE 1.2   Basic computer architecture.

1.3   PRACTICE: HOW TO CHECK WHAT KIND OF CPU YOUR COMPUTER HAS

Most readers will have access to computers, if not you can go to the local library to use the computer. You can easily find what kind of CPU (the circle in Figure 1.4) and memory device (rectangular in Figure 1.4) the computer has without removing the cover of the computer. Please follow the following steps (the following description and figures are made by *Windows* [*Windows 11*], not *IOS*).

First, type 'settings' (Figure 1.5), and then press 'enter'. Then, you should see Figure 1.6. Scroll down, and you can find the 'About' button (Figure 1.6).

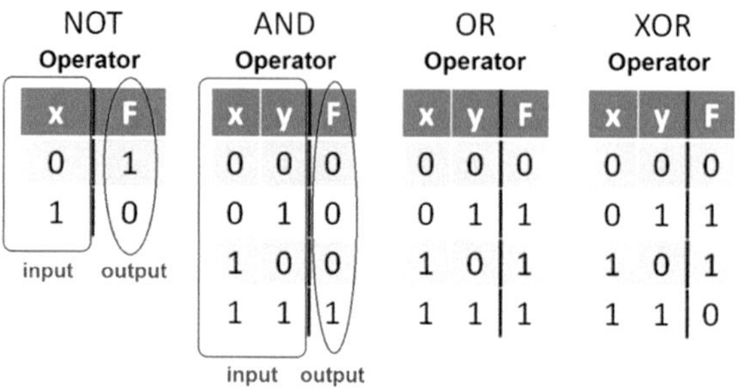

NOT		AND			OR			XOR		
\multicolumn 2 c Operator		\multicolumn 3 c Operator			\multicolumn 3 c Operator			\multicolumn 3 c Operator		

x	F	x	y	F	x	y	F	x	y	F
0	1	0	0	0	0	0	0	0	0	0
1	0	0	1	0	0	1	1	0	1	1
		1	0	0	1	0	1	1	0	1
		1	1	1	1	1	1	1	1	0

input   output

input   output

FIGURE 1.3 Four basic components of logic operation.

Click the 'About' button. Then, you should find a page somewhat similar to that in Figure 1.7.

By analyzing the page similar to Figure 1.7, you can easily discover what kind of CPU and memory device the computer has. As emphasized by the rectangle in Figure 1.7, this computer has a CPU (processor) made by Intel, i7-10800F, with 2.90 GHz. In addition, this computer has 32.0 GB main memory (RAM).

For clarity, note that regarding the memory of a computer, the computer has two kinds of memory: The main memory (e.g., RAM) and auxiliary memory (e.g., hard disk, CD, DVD, SSD). The illustrated computer has 32 GB main memory (RAM) and 512 GB auxiliary memory (SSD). It can save music files, game files, and document files up to 512 GB. The main memory (RAM) acts as a bridge between the auxiliary memory (hard disk, CD, DVD, SSD) and the CPU. A more detailed explanation on the differences between main memory and auxiliary memory can be found in References [5–8].

1.4   VON NEUMAN ARCHITECTURE

So far, we have learned that a computer is composed of an input unit, CPU, memory unit, and output unit. This concept in dealing with computer architecture is called 'Von Neuman architecture' [9]. The input unit includes the keyboard, mouse, and scanner, while the output unit includes the monitor, printer, and speaker. Importantly, as shown in Figure 1.8, the CPU has the function of arithmetic operation (addition, deduction) and logic operation (NOT, OR, AND). The memory devices help the computer to save information.

In this book, the most recent state-of-the-art technology is explained in detail, especially for CPU and memory. In both CPU and memory, a negative capacitance field effect transistor (NCFET) plays a key role in designing high-performance CPU and memory. To understand the concept of NCFET, we need to learn the concept of the circuit and transistors. A recently developed computer has more than 0.1 billion circuits, and one circuit has about 2–20 transistors. Therefore, this computer could have more than 1 billion transistors.

For a bottom-up approach, in this book, 'transistor' is explained first, and then 'circuit' is explained later. Figure 1.9 briefly shows the basic structure of modern

FIGURE 1.4 The mainboard of a computer.

FIGURE 1.5 First step for hands-on practice: checking the CPU type of a computer.

transistors, which is usually a metal oxide semiconductor field effect transistor (MOSFET). In this MOSFET, an electric voltage can be applied to the source, gate, and drain region (Figure 1.9) through the metal line. Normally, the source has 0 V, and the drain has 0.6 V.

However, the gate voltage varies, depending on the situation. When you want to turn this transistor (MOSFET) on, then you can apply a gate voltage of 0.6 V. Then, electric current will flow in the MOSFET. On the contrary, if you want to turn the

FIGURE 1.6 Second step for hands-on practice: checking the CPU type of a computer.

transistor (MOSFET) off, then you might apply a gate voltage of 0 V. Then, electric current will not flow through the transistor (MOSFET).

Specifically, when the transistor is on (gate voltage = drain voltage = 0.6 V and source voltage = 0 V), lots of electrons move from the source (0 V) to the drain (0.6 V).

Then, the current will flow from the drain to the source (the direction of current is the opposite of the direction of electron movement).

Figure 1.10 briefly shows how the structure of MOSFET has evolved [10–16]. When MOSFET was first developed in 1960, the structure of planar MOSFET (Figure 1.10) was used. This planar MOSFET (Figure 1.10) structure is the same structure as in Figure 1.9. TAs time has passed, the structure of this planar MOSFET evolved into the double gate MOSFET (DG-MOSFET), FinFET, and GAA MOSFET (= nanosheet [NS] MOSFET), step by step. Specifically, the advanced structure has better controllability. That is to say, when the transistor is off, the advanced structure consumes low power. Therefore, a modern computer made using an advanced structure consumes low power, thereby saving battery power. To sum up, the future of transistors requires: (1) high performance (high on-state current), (2) low power consumption (low off-state current), and (3) fast switching speed (low delay). By incorporating the NCFET technique, these three aims for future transistors could be simultaneously achieved [17–20].

FIGURE 1.7 Third step for hands-on practice: checking the CPU type of a computer.

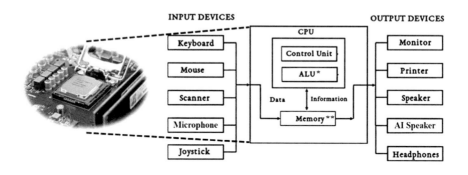

ALU stands for Arithmetic Logic Unit, which plays important roles as adder, multiplier, logic operator
** *Memory consists of main memory (RAM) and auxiliary memory (Hard disk, CD, DVD, SSD)*

FIGURE 1.8 Von Neumann architecture with a CPU.

The title of this book (*Negative Capacitance Field Effect Transistors* [NCFETs]) describes the state of art technology which has been applied to all these structures. By applying this NCFET technology to the transistor, the performance of the transistor can be significantly boosted, and the power consumption of the CPU (during the off-state, when the computer or portable electronic device is switched off) could

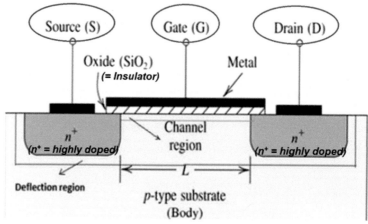

FIGURE 1.9   The basic structure of MOSFET with three contacts (source, gate, drain).

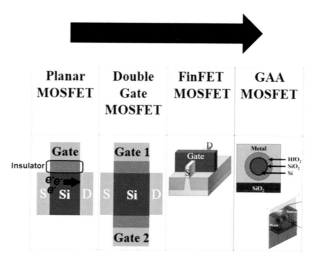

FIGURE 1.10   Structural evolution of MOSFET from planar MOSFET (initial structure) to GAA MOSFET (recent advanced structure).

be decreased at the same time. In the next chapter, the basic operation principle of NCFET is described step by step.

1.5   CONCLUSION

So far, we have learned about the basic structure of modern computers, namely Von Neumann architecture. Most computers consist of an input unit, central processing unit (CPU), memory unit, and output unit. Among these, the CPU plays a key role in the computer, especially with logic operations (NOT, OR, AND) and basic arithmetic operations (addition, deduction). The CPU can consist of more than 0.1 billion circuits, and one circuit normally consists of about 2–20 transistors. Therefore, one CPU can have more than 1 billion transistors, and so the basic design of transistors is paramount. In this regard, by incorporating NCFET technology (which is described later in this book), the three main goals of future transistor design—high performance, low power, high switching speed—could be simultaneously achieved.

ACKNOWLEDGMENTS

The authors would like to their express heartfelt appreciation to various previous researches and image sites (e.g., Pixabay) which have inspired the authors to have creative thinking. Some figures have been re-edited utilizing free non-copyright photos (Pixabay). We have referred to and cited various sources in this chapter.

REFERENCES

1. H. Ando et al. (2003) A 1.3 GHz fifth generation SPARC64 microprocessor. *2003 IEEE International Solid-State Circuits Conference, 2003. Digest of Technical Papers. ISSCC*, vol.1, pp. 246–491. San Francisco, CA, USA, 2003.
2. B. A. Kuncoro and B. H. Iswanto (2015) TF-IDF method in ranking keywords of Instagram users' image captions. *2015 International Conference on Information Technology Systems and Innovation (ICITSI)*, pp. 1–5. Bandung, Indonesia.
3. G. E. Moore (2003) No exponential is forever: but "Forever" can be delayed! [semiconductor industry]. *2003 IEEE International Solid-State Circuits Conference, 2003. Digest of Technical Papers. ISSCC*, vol. 1, pp. 20–23. San Francisco, CA, USA.
4. P. Gargini (2017) Roadmap evolution: From NTRS to ITRS, from ITRS 2.0 to IRDS. *2017 Fifth Berkeley Symposium on Energy Efficient Electronic Systems & Steep Transistors Workshop (E3S)*, pp. 1–62. Berkeley, CA, USA.
5. M. Shaafiee, R. Logeswaran and A. Seddon (2017) Overcoming the limitations of von Neumann architecture in big data systems. *2017 7th International Conference on Cloud Computing, Data Science & Engineering – Confluence*, , pp. 199–203. Noida, India.
6. R. Nair (2015) Evolution of Memory Architecture. *Proceedings of the IEEE* 103(8): 1331–1345.
7. A. S. Nayak and M. Vijayalakshmi (2013) Teaching Computer System Design and Architecture course—An experience. *2013 IEEE International Conference in MOOC, Innovation and Technology in Education (MITE)*, pp. 21–25. Jaipur, India,
8. C. Barnes, P. Vaidya and J. J. Lee (2009) An XML-based ADL framework for automatic generation of multithreaded computer architecture simulators. *IEEE Computer Architecture Letters* 8(1): 13–16.
9. L. Koskinen, J. Tissari, J. Teittinen, E. Lehtonen, M. Laiho and J. H. Poikonen (2016) A performance case-study on memristive computing-in-memory versus Von Neumann architecture. In: *2016 Data Compression Conference (DCC)*, pp. 613–613. Snowbird, UT, USA.

10. Y. S. Song, S. Kim, J. H. Kim, G. Kim, J.-H. Lee and W. Y. Choi (2023) Enhancement of thermal characteristics and on-current in GAA MOSFET by utilizing Al_2O_3-based dual-κ spacer structure. *IEEE Transactions on Electron Devices* 70(1): 343–348.

11. S. Tayal et al. (2022) Incorporating bottom-up approach into device/circuit co-design for SRAM-based cache memory applications. *IEEE Transactions on Electron Devices* 69(11): 6127–6132.

12. K. Y. Kim, Y. S. Song, G. Kim, S. Kim and J. H. Kim (2022) Reliable high-voltage drain-extended FinFET with thermoelectric improvement. *IEEE Transactions on Electron Devices* 69(11): 5985–5990.

13. Y. S. Song, S. Tayal, S. B. Rahi, J. H. Kim, A. K. Upadhyay and B.-G. Park (2022) Thermal-aware IC chip design by combining high thermal conductivity materials and GAA MOSFET. In: *2022 5th International Conference on Circuits, Systems and Simulation (ICCSS)*, pp. 135–140. Nanjing, China.

14. S. J. Kang, J. H. Kim, Y. S. Song, S. Go and S. Kim (2022) Investigation of self-heating effects in vertically stacked GAA MOSFET with wrap-around contact. *IEEE Transactions on Electron Devices* 69(3): 910–914.

15. Y. S. Song, J. H. Kim, G. Kim, H.-M. Kim, S. Kim and B.-G. Park (2021) Improvement in self-heating characteristic by incorporating hetero-gate-dielectric in gate-all-around MOSFETs. *IEEE Journal of the Electron Devices Society* 9: 36–41.

16. J.-P. Colinge (2003) The evolution of silicon-on-insulator MOSFETs. In: *International Semiconductor Device Research Symposium, 2003*, pp. 354–355. Washington, DC, USA.

17. H. Agarwal et al. (2019) Proposal for capacitance matching in negative capacitance field-effect transistors. *IEEE Electron Device Letters* 40(3): 463–466.

18. J. Li et al. (2017) Correlation of gate capacitance with drive current and transconductance in negative capacitance Ge PFETs. *IEEE Electron Device Letters* 38(10): 1500–1503.

19. S. Durukan, O. Palamutçuoğullari and A. E. Yilmaz (2022) CMOS negative impedance converter circuit with the elimination of parasitic gate-source capacitance. In: *2022 Microwave Mediterranean Symposium (MMS)*, pp. 1–5. Pizzo Calabro, Italy.

20. T. Dutta, V. Georgiev and A. Asenov (2018) Random discrete dopant induced variability in negative capacitance transistors. In: *2018 Joint International EUROSOI Workshop and International Conference on Ultimate Integration on Silicon (EUROSOI-ULIS)*, pp. 1–4. Granada, Spain.

2 Technical Demands of Low-Power Electronics

Soha Maqbool Bhat[1], Pooran Singh[1], Ramakant Yadav[1], Shiromani Balmukund Rahi[2], Billel Smaani[3], Abhishek Kumar Upadhyay[4], Young Suh Song[5]

[1]Department of Electrical & Electronics Engineering, Mahindra University, Hyderabad, India
[2]Department of Electrical Engineering, Indian Institute of Technology, Kanpur 208016, India
[3]Centre Universitaire Abdelhafid Boussouf – Mila, Mila 43000, Algeria
[4]X-FAB Semiconductor foundries, 99097 Erfurt, Germany
[5]Korea Military Academy, Seoul

2.1 INTRODUCTION

Energy saving is a most promising sector of research and development. Nowadays, energy saving or low power is becoming a critical challenge that is most important for the future of humanity. Semiconductor science and technology play a continuous lead role in energy sectors. The development of the first semiconductor device named BJT is the first example. This device has almost totally replaced vacuum tube technology. Exiting leakage currents in BJT have become its main negative factor. Research and development have resolved a solution for BJT technology in terms of field effect technology. MOSFET (metal oxide semiconductor field effect transistor) is one of the most popular candidates in field effect technology. This semiconductor device has played the lead role in the development of low-power circuits and systems. It has completely changed the thinking and lifestyles of humans. Laptops, smart phones, and smartwatches are some of its common uses, and the most popular development is based on MOSFET technology. The bulk of MOSFET-based devices have followed the rules and regulations timely published by ITRS and other agencies such as IDEM and Moore's law. The demand for low power is continuously increasing, and power supply scaling has been the main guiding rule for MOSFETs.

This chapter is an overview presenting the technical demand for low-power applications. The following section focuses on details about the technical requirements of low-power research and development. In this chapter, we briefly mention some of the most relevant and emerging semiconductor field effect devices and their applications for low-power very large-scale integration (VLSI).

2.2 TECHNICAL DEMAND INVESTIGATION FOR LOW-POWER USES

In electrical engineering, the power dissipation of an electronic circuit and system during a particular time interval of a circuit element can be measured by relation (2.1), written as

$$E = \int P(t)\,dt \tag{2.1}$$

In Equation (2.1), E is energy and P is power. The power dissipation of an integrated circuit and system can be represented by Equation (2.2)

$$P = P_{switch} + P_{sc} + P_{off} \tag{2.2}$$

Equation (2.2) shows that the power dissipation roughly contains three components, P_{switch} is the switching power, P_{sc} is the power corresponding to the simultaneous and short-time *on* state of the two MOSFETs, due to their non-ideal behavior, and P_{off} is the static power dissipation. In a VLSI circuit and system, design engineers reduce the switching power P_{switch} at a high level using ultra-scaled MOSFET structures. This is possible due to continuous scaling of conventional MOSFET, as estimated by Equation (2.3), where α is the switching activity, f the frequency, C the capacitance, and V_{DD} the voltage applied

$$P_{switch} = \alpha f C V_{DD}^2 \tag{2.3}$$

The scaling of the MOSFET reduces the switching power but increases the off-state leakage power dissipation. This power is directly connected to the leakage current of MOSFET denoted as I_{OFF}, calculated by the following well-known equation.

$$E_{switch} = C V_{DD}^2 \tag{2.4}$$

In CMOS, both the energy (E) and the power (P) dissipated are related to the square of the applied voltage, V_{DD}.

2.3 CHALLENGES OF CONVENTIONAL CMOS TECHNOLOGY

One of the most important key parameters is the subthreshold-swing (SS), which represents the basic performance of transistors. For example, a transistor that has lower SS will have a low off-current and high on-current, thereby realizing both low power consumption and high performance, at the same time. However, there has been a fundamental limitation regarding SS. That is to say, conventionally, SS has been difficult to have lower than 60 mV/dec, because of the physical mechanism of electrical current in the transistor (namely, thermionic emission).

The low-power R&D block point in term of technology is known as 'Boltzmann's tyranny.' For MOSFET-based technology, this key term is formulated as

$$SS = \left(\frac{d\left(log_{10}I_{DS}\right)}{dV_{GS}}\right)^{-1} = \frac{dV_{GS}}{d\psi_S} \times \frac{d\psi_S}{d\left(log_{10}I_{DS}\right)} = \left(1+\frac{C_d}{C_{ox}}\right) \times \frac{d\psi_S}{d\left(log_{10}I_{DS}\right)}$$

The numeric value of SS calculated for 300 K is:

$$SS = 2.3 \times 25.8 \frac{mV}{decade} \approx 60 \frac{mV}{decade}$$

In the developed low-power series , the semiconductor has overshot the limitations of conventional MOSFET. The solution of subthreshold slope limitations has been identified in terms of steep subthreshold FETs such as tunnel FETs and NCFETs. These devices have a lower steep subthreshold than 60 mV/decade.

Figure 2.1 shows a schematic of the conventional MOSFET and the effect of scaling on the device dimensions. It has three metal contacts: source, gate, and drain. The electric voltage can be applied to these three contacts. By applying a high voltage to the gate, the transistor can be turned on. On the other hand, by applying 0 V to the gate, the transistor can be turned off.

Figure 2.2 shows that continuous power scaling is impossible because leakage power has become the dominant consumer of power for the smaller 90 nm technology node. To overshoot power scaling limitation, device researchers have suggested and developed several types of field effect devices, namely tunnel FETs and negative capacitance FETs, which are the most effective developments. TFET technology replaces classical transport in the case of conventional MOSFET with quantum-mechanical band-to-band (BtB). BtB tunneling results in considerable and remarkable enhancements in subthreshold slope (SS) and power consumption characteristics, far beyond those of the standard CMOS technology. With the CMOS process compatibility feature, tunnel FETs have been proven to be prominent structures to meet the requirements of low-power nano-devices because of their lower inverse sub-threshold slope (SS) and low OFF current (Figure 2.3) [14–16].

In parallel with the development of tunnel FET technology, semiconductor players adopted the negative capacitance feature found in ferroelectric materials identified by

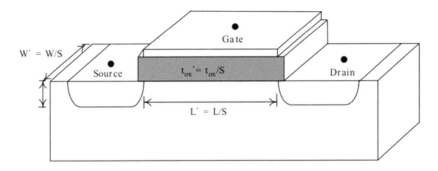

FIGURE 2.1   Scaling of MOSFET. Here S > 1 is the scaling factor.

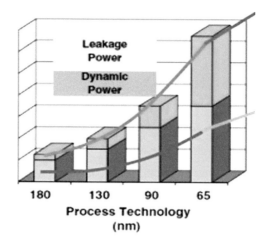

FIGURE 2.2 Power consumption for different technology nodes.

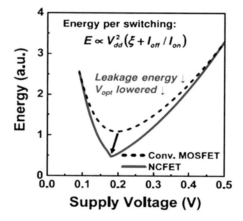

FIGURE 2.3 Technical requirements of low-power research and developments [2].

Suprio Dutta's group. Negative capacitance-based technology has no transport phenomena issues like tunnel FETs. Both tunnel FET and NC FET devices have steep subthresholds and operate at a supply voltage lower than 0.7 V. The low supply operating feature clearly illustrates that both of these FET candidates are strong semiconductor players for low-power circuits and systems design.

Figure 2.4 shows the supply voltage and EOT scaling in the MOSFET technologies [17]. The EOT limit defined by the SiO_2 layer between the Si channel and HfO_2 material is indicated by the black dashed line. The minimum supply voltage caused by the Boltzmann limit is illustrated by the red dashed lines in Figure 2.4.

The CMOS devices have accomplished remarkable progress with higher speed and lower power for more than 50 years. Table 2.1 shows the device trends in the past (based on IEDM) and future (ITRS) of high-performance (HP) devices. The steep reductions in time delay and energy consumption are achieved by device scaling, which remarkably contributes to the innovations of IT technologies.

2.4   QUANTUM DOT CELLULAR AUTOMATA TECHNOLOGY

Quantum dot cellular automata (QCA) is the most promising technology which incorporates a shift from transistorized designs to the design of digital circuits using quantum dots or metal islands. The vital component of QCA circuits is a QCA cell made up of four quantum dots [18]. These quantum dots are 18×18 nm in size and are located at the cell corners. In the adjacent quantum dots, two free electrons reside. These electrons are not able to move among the cells but can tunnel among the dots. There are two polarization states of the electrons. They will be either '0' or '1'. These polarization states are due to the Coulombic repulsion between the cells. The information flow from a cell to an empty cell is due to the Coulombic repulsion [19].

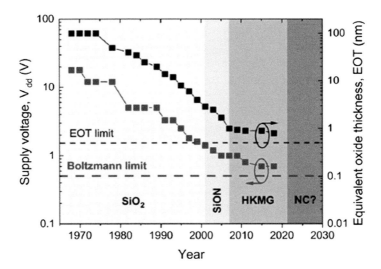

FIGURE 2.4   Historical trend of the supply voltage V_{dd} and equivalent oxide thickness (EOT) scaling in conventional MOSFET technologies.

TABLE 2.1
International technology roadmap for semiconductors

Year production		2021	2022	2023	2024	2025	2026
L	L_g	10.2	9.2	8.2	7.4	6.6	5.9
S	EOT (nm)	8.0	0.75	0.71	0.68	0.64	0.6
T	V_{dd} (V)	0.65	0.63	0.61	0.59	0.56	0.54
P	L_g (nm)	9.7	8.9	8.1	7.4	6.6	5.9
H	EOT (nm)	0.59	0.56	0.53	0.5	0.47	0.45
P	V_{dd} (V)	0.66	0.64	0.62	0.61	0.59	0.57

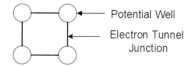

FIGURE 2.5   Basic QCA cell.

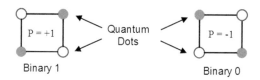

FIGURE 2.6   Polarizations and representations of binary 1 and binary 0.

FIGURE 2.7   QCA wires.

Figure 2.5 shows the basic component of QCA circuits which is known as a QCA cell and Figure 2.6 shows the polarization states of the cell.

The QCA wire is a chain of QCA cells which helps in the flow of information. There are two types of QCA wires: 90° wire and 45° wire, as shown in Figure 2.7.

Another vital component in QCA circuits is a three-input majority voter gate, as shown in Figure 2.8. The three-input majority gate can be logically implemented as an OR gate and an AND gate by manipulating the inputs. To execute the majority voter gate as an OR gate fix one of the three inputs as logic '0' and to implement the AND gate fix it as logic '1'.

The inverter is another vital component in QCA circuits. The QCA designs of inverters existing till date are shown in Figure 2.9.

The clocking in QCA circuits is for the proper flow of information and supplies power to reform the signal power dissipation to the setting. There are four clock levels in QCA: switch, hold, release, and relax. as shown in Figure 2.10. In the switch phase, the inter dot barrier is enhanced and the cell switches its polarity. In the hold phase, the cell keeps its polarity attained in the switch phase. The cell loses its polarity during the release phase. The cell attains null polarity at the relax phase and electrons transfer openly in the cell [18,19]. Moreover, these phases are represented by four distinct colors. The switch phase is represented by green, the hold phase by magenta, the release phase by blue, and the relax phase by white. The cells in a particular zone are managed by the identical clock signal and form a sub-array.

A non-adiabatic clocking scheme can provide a worst-case power estimate in QCA cells. Using QCA circuits, we determine the maximum power dissipated. QCA cells are analyzed using quantum mechanical calculations to determine their power dissipation and polarization. $|1>$ and $|0>$ are the eigenstates of a QCA cell.

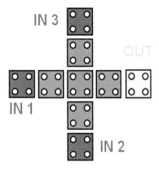

FIGURE 2.8 Three input majority voter gate in QCA.

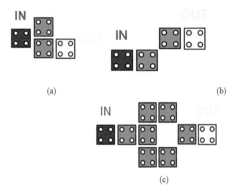

FIGURE 2.9 QCA inverter configurations: (a) rotated cell, (b) half-cell displaced, (c) robust.

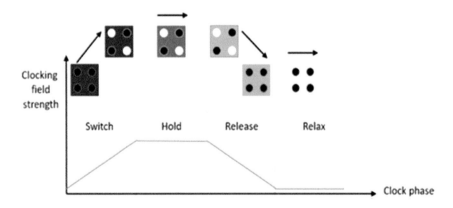

FIGURE 2.10 The clocking zones in QCA [18,19].

The Hamiltonian is the Schrodinger equation that describes a wave function at a given point in time. Schrodinger wave equations are used to equate Hamilton's equation, which is a 2×2 matrix. In matrix form, Equations (2.3) and (2.4) represent Hamiltonian equations.

$$Hj = \begin{bmatrix} -Ek \div 2\sum Fj,m\,Pm & -\gamma \\ -\gamma & Ek \div 2\sum Fj,m\,Pm \end{bmatrix} \tag{2.3}$$

$$Hj = \begin{bmatrix} -EkP \div 2 & -\gamma \\ -\gamma & EkP \div 2 \end{bmatrix} \tag{2.4}$$

Here, γ represents the tunneling energy between two polarized states and E_k is the kink energy. The kink energy is defined as the difference between two horizontally adjacent polarized cells whose polarizations are either the same or opposite. Fj,m is the geometric factor that varies with j-th and m-th cells in distance and orientation, and Pm is the polarization of the cell considered. The flow of energy can be calculated from the Hamiltonian equations and density matrix. As the clock level rises or barriers are raised, energy flows into the cell, and as the clock level drops or barriers are lowered, it returns to the clock. The energy must be zero in the steady state. The power flow can be calculated by differentiating the energy equation as given in References [20,21]. The upper limit for energy dissipated by a QCA cell during a switching event is given as:

$$Ediss = \left[2\gamma new\,/\,Ek\left(\frac{Po}{Pold}\gamma old - \frac{Pn}{Pnew}\gamma new \right) \right. \\ \left. + \left[Ek\,\frac{Pnew}{2}(Pn - Po) \right] \right] \tag{2.5}$$

If t is the energy relaxation time then the power dissipation can be calculated as:

$$Pdiss = 1/t\left[2\gamma new/Ek\left(\frac{Po}{Pold}\gamma old - \frac{Pn}{Pnew}\gamma new \right) \right. \\ \left. + Ek\,\frac{Pnew}{2}(Pn - Po) \right] \tag{2.6}$$

Let P_n and P_o be the input and output cell polarization, P_{old} and P_{new} are the polarization states before and after a switching event, and γ_{new} and γ_{old} are the clock energy during the switching event.

Figure 2.11 shows an example of a D-latch-based memory cell and Figure 2.12 shows the energy dissipation maps at different tunneling levels that are calculated using QCAPro tool as in Reference [22]. The darker region shows more energy dissipation and the lighter region shows less dissipation of energy. Therefore, the memory cell dissipates less energy at the 0.5 Ek level as compared to the 1.0 and 1.5 Ek levels.

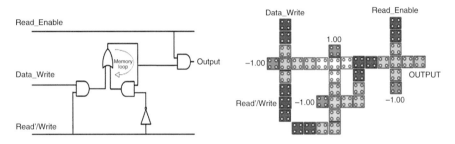

FIGURE 2.11 (a) Representation of D-latch based memory cell, (b) implementation of memory cell in QCA [22].

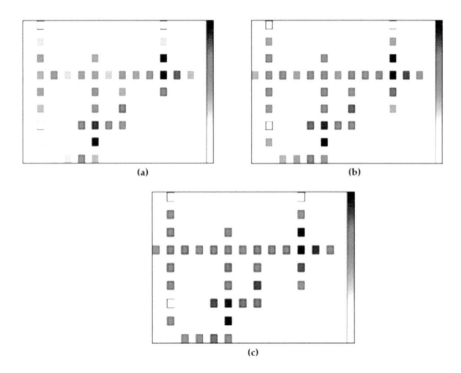

FIGURE 2.12 Energy dissipation of memory cell at (a) 0.5 *Ek* (b) 1.0 *Ek*, and (c) 1.5 *Ek* levels [22].

2.5 NEGATIVE CAPACITANCE FIELD EFFECT TRANSISTORS (NC-FETS)

CMOS technology faces fundamental limitations that may be overcome by negative capacitance field effect transistors (NCFETs). As NCFETs feature a ferroelectric layer within their gates, which amplifies voltage internally, they can operate at a lower voltage and still achieve high performance with comparatively little energy [23]. Their potential for subthreshold swing (SS) devices below 60 mV/decade has attracted considerable interest over the past 5 years. In addition to their capability to scale to the next technological node, sub-60-mV/decade devices also limit device power dissipation, enabling transistor density and performance to be increased. By replacing the gate oxide with a ferroelectric dielectric stack, NCFETs improve device performance through 'negative capacitance' [24].

As reported in References [25–29], a metal-ferroelectric metal–insulator semiconductor (MFMIS) structure is used in short-channel NCFET devices, which require a metal layer between the ferroelectric and dielectric layers. The MFMIS structure has a higher implementation cost and is less practical because of biasing issues [30]. In Reference [31], a device without the metal layer, the metal–ferroelectric insulator semiconductor (MFIS) structure, is reported. It has been shown in References [24,32,33] that hysteresis-free NCFETs are possible if a greater ferroelectric capacitance than the field effect transistor (FET) capacitance exists.

Ferroelectric depolarization is responsible for passively boosting the voltage across the internal gate of NCFETs. Under identical operating conditions, the surface potential of NCFETs is higher than that of conventional MOSFETs [34–39]. Unlike TFETs, this device technology adds amplification to the internal voltage but does not change the charge transport. This type of device employs non-linearity in the gate insulator [42–45]. The NC effect arising from the negative coefficient around the metastable state can be stabilized by connecting a paraelectric capacitor in series with a ferroelectric [44–51].

A 3D view of an NCFET is shown in Figure 2.13. The equivalent symbolic representation of an NCFET is shown in Figure 2.14.

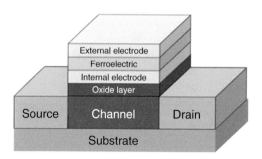

FIGURE 2.13   Schematic of NCFET [42].

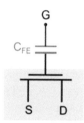

FIGURE 2.14   Symbol of NCFET [42].

TABLE 2.2
Performance parameters NCFET and MOSFET at 1 V

	NCFET		Baseline MOSFET	
	N-type	**P-type**	**N-type**	**P-type**
SS (mV/decade)	53.92	58.96	73.24	79.58
I_{on} (mA/μm)	1.61	0.94	1.40	0.81
I_{off} (mA/μm)	0.0084	0.0076	0.28	0.26
On/off ration	1.91×10^8	1.22×10^8	5×10^6	3.18×10^6

TABLE 2.3
Power comparison of SRAM

Power consumption	MOSFET	NCFET	QCA [22]
P_{Total}	494.7 pW	85.3 pW	9.61 pW

An NCFET-based 6T SRAM design has been investigated in Reference [43]. It has been reported as an excellent choice over CMOS-based 6T SRAM because of the high yield estimation at near the threshold voltage. A significant improvement has been observed in the design using NCFETs. Table 2.2 presents the performance of a 6T SRAM based on NCFETs and baseline MOSFETS at V_{DD} equal to 1 V [43].

Table 2.3 shows the power comparison of NCFET-, MOSFET-, and QCA-based SRAMs. The QCA-based SRAM dissipates less power as compared to the NCFET- and MOSFET-based SRAM cells. The QCA technology has some limitations such as the fabrication of QCA devices is still under testing and these devices cannot operate above room temperature. Therefore, NCFET is a more promising technology.

Table 2.4 shows the temperature sensitivity of a negative capacitance FET. As shown in this table, at room temperature (i.e., 300 K), the SS value is close to 60 mV/decade. This illustrates the theoretical compatibility of negative capacitance with ideal limitations of classical CMOS technology. The low off current, lower SS, and equivalent on-state current with existing FETs like classical NCFET, NC-tunnel FET, NC-FinFET, and NC-GAA could help future development of an ultra-low-power circuit and system [32–53].

Table 2.5 shows the design facility for optimized NCFETs of negative capacitance FETs. As shown in this table, the appropriate selection of ferroelectric materials can help to optimize the device advantages. The internal voltage amplification of ferroelectric-based devices and circuit helps to reduce the power consumption and in switching energy.

TABLE 2.4
Impact of temperature on SS of NCFET

S. No.	Temperature (K)	SS (mV/decade)	Remarks
1	280	56.5	
2	**300**	**59.0< 60.0**	
3	320	64.7	
4	340	68.2	
5	360	71.2	

Temperature   SS

Xiao, Y.G., Tang, M.H., Li, J.C., Cheng, C.P., Jiang, B., Cai, H.Q., ... & Gu, X.C. (2012) Temperature effect on electrical characteristics of negative capacitance ferroelectric field-effect transistors. *Applied Physics Letters* 100(8): 083508.

TABLE 2.5
Impact of ferroelectric thickness (T_{FE}) on SS of NCFET

S.N.	T_{FE} (nm)	SS (mV/decade)	Remarks
1	0	88	
2	40	82	
3	80	75	
4	120	68	

T_{FE}(nm)   SS (mV/decade)

2.6   CONCLUSION

Future nano-electronic circuits and systems face a great challenge in dissipating power in switching devices. This chapter reviews ultra-low leakage, ultra-low voltage, stepper inverse subthreshold slope, and ultra-low-power devices for the Internet of Things. Among the emerging technologies, NCFETs are capable of going beyond the fundamental limits of conventional CMOS technology. Despite reducing power computation and switching energy, NCFETs maintain performance. It is expected that NCFET technology will bypass the limitations of current CMOS technology, according to available scientific information on semiconductor device technology and that it will contribute greatly to future growth.

REFERENCES

1. Lent, C.S., Tougaw, P.D., Porod, W., Bernstein, G.H. (1993) Quantum cellular automata. *Nanotechnology* 4: 49.
2. Lu, Y., Liu, M., Lent, C. (2007) Molecular quantum-dot cellular automata: From molecular structure to circuit dynamics. *Journal of Applied Physics* 102: 034311.
3. Lu, Y., Liu, M., Lent, C. (2006) Molecular electronics-from structure to circuit dynamics. In: *Proceedings of Sixth IEEE Conference on Nanotechnology*, pp. 62–65.
4. Frost, S.E., Rodrigues, A.F., Janiszewski, A.W., Rausch, R.T., Kogge, P.M. (2002) Memory in motion: A study of storage structures in QCA. In: *Proceedings of First Workshop on Non-Silicon Computing*, pp. 1–8.
5. Blair, E.P., Yost, E., Lent, C.S. (2010) Power dissipation in clocking wires for clocked molecular quantum-dot cellular automata. *Journal of Computational Electronics* 9: 49–55.
6. Ahmed, S., Baba, M.I., Bhat, S.M., Manzoor, I., Nafees, N., Ko, S.-B. (2020) Design of reversible universal and multifunctional gate-based 1-bit full adder and full subtractor in quantum-dot cellular automata nanocomputing. *Journal of Nanophotonics* 14: 036002.
7. Nafees, N., Ahmed, S., Kakkar, V., Bahar, A.N., Wahid, K.A., Otsuki, A. (2022) QCA-based PIPO and SIPO shift registers using cost-optimized and energy-efficient D flip flop. *Electronics* 11: 3237.
8. Almatrood, A., George, A.K., Singh, H. (2021) Low-power multiplexer structures targeting efficient QCA nanotechnology circuit designs. *Electronics* 10: 1885.
9. Safoev, N., Jeon, J.-C. (2020) Design and evaluation of cell interaction based vedic multiplier using quantum-dot cellular automata. *Electronics* 9: 1036.
10. Yan, A., Liu, R., Huang, Z., Girard, P., Wen, X. (2022) Designs of level-sensitive T flip-flops and polar encoders based on two XOR/XNOR gates. *Electronics* 11: 1658.
11. Bahar, A.N., Wahid, K.A. (2020) Design of an efficient N× N butterfly switching network in quantum-dot cellular automata (QCA). *IEEE Transactions on Nanotechnology* 19: 147–155.
12. Seyedi, S., Pourghebleh, B., Jafari Navimipour, N. (2022) A new coplanar design of a 4-bit ripple carry adder based on quantum-dot cellular automata technology. *IET Circuits, Devices and Systems* 16: 64–70.
13. Enayati, M., Rezai, A., Karimi, A. (2021) Efficient circuit design for content-addressable memory in quantum-dot cellular automata technology. *SN Applied Sciences* 3: 1–10.
14. Yadav, R., Dan, S.S., Vidhyadharan, S., Hariprasad, S. (2021) Suppression of ambipolar behavior and simultaneous improvement in RF performance of gate-overlap tunnelfield effect transistor (GOTFET) devices. *Silicon* 13: 1185–1197.
15. Yadav, R., Dan, S.S., Vidhyadharan, S., Hariprasad, S. (2020) Innovative multi-threshold gate-overlap tunnel FET (GOTFET) devices for superior ultra-low power digital, ternary and analog circuits at 45-nm technology node. *Journal of Computational Electronics* 19(1): 291.
16. Yadav, R., Dan, S.S., Hariprasad, S. (2022) Low and high Vt GOTFET devices outperform standard CMOS technology in ternary logic applications. *IETE Technical Review* 39: 1114–1123.
17. Hoffmann, M., Slesazeck, S., Mikolajick, T. (2021) Progress and future prospects of negative capacitance electronics: A materials perspective. *APL Materials* 9: 020902.

18. Bilal, B., Ahmed, S., Kakkar, V. (2018) An insight into beyond CMOS next generation computing using quantum-dot cellular automata nanotechnology. *International Journal of Engineering and Manufacturing* 11: 25.

19. Bilal, B., Ahmed, S., Kakkar, V. (2018) Quantum dot cellular automata: A new paradigm for digital design. *International Journal of Nanoelectronics and Materials* 11: 87–98.

20. Lent, C.S., Timler, J. (2002) Power gain and dissipation in quantum cellular automata. *Journal of Applied Physics* 91: 823–831.

21. Ganesh, E.N. (2014) Power analysis of quantum cellular automata circuits. In: *2nd International Conference on Nanomaterials and Technologies*, pp. 381–394.

22. Bhat, S.M., Ahmed, S., Bahar, A.N., Wahid, K.A., Otsuki, A., Singh, P. (2023) Design of cost-efficient SRAM cell in quantum dot cellular automata technology. *Electronics* 12: 367.

23. Amrouch, H. van Santen, V.M., Pahwa, G., Chauhan, Y., Henkel, J. (2020) NCFET to rescue technology scaling: Opportunities and challenges. In: *2020 25th Asia and South Pacific Design Automation Conference (ASP-DAC)*, pp. 637–644. Beijing, China.

24. Salahuddin, S., Datta, S. (2008) Use of negative capacitance to provide voltage amplification for low power nanoscale devices. *Nanoscale Letters* 8(2): 405–410.

25. Seo, J., Lee, J., Shin, M. (2017) Analysis of drain-induced barrier rising in short-channel negative-capacitance FETs and its applications. *IEEE Transactions on Electron Devices* 64(4): 1793–1798.

26. Huang, S.-E., Yu, C.-L., Su, P. (2018) Investigation of fin-width sensitivity of threshold voltage for InGaAs and Si negative-capacitance FinFETs considering quantum-confinement effect. *IEEE Transactions on Electron Devices* 66(6): 2538–2543.

27. Agarwal, H. et al. (2018) Designing 0.5 V 5-nm HP and 0.23 V 5-nm LP NC-FinFETs with improved IOFF sensitivity in presence of parasitic capacitance. *IEEE Transactions on Electron Devices* 65(3): 1211–1216.

28. Dutta, T., Pahwa, G., Agarwal, A., Chauhan, Y.S. (2018) Impact of process variations on negative capacitance FinFET devices and circuits. *IEEE Electron Device Letters* 39(1): 147–150.

29. Mehta, H., Kaur, H. (2019) Study on impact of parasitic capacitance on performance of graded channel negative capacitance SOI FET at high temperature. *IEEE Transactions on Electron Devices* 66(7): 2904–2909.

30. Singh, R. et al. (2018) Evaluation of 10-nm bulk FinFET RF performance—Conventional versus NC-FinFET. *IEEE Electron Device Letters* 39(8): 1246–1249.

31. Khan, A.I., Radhakrishna, U., Chatterjee, K., Salahuddin, S., Antoniadis, D.A. (2016) Negative capacitance behavior in a leaky ferroelectric. *IEEE Transactions on Electron Devices* 63(11): 4416–4422.

32. Cam, T. et al. (2020) Sustained benefits of NCFETs under extreme scaling to the end of the IRDS. *IEEE Transactions on Electron Devices* 67(9): 3843–3851.

33. Lin, C.-I., Khan, A.I., Salahuddin, S., Hu, C. (2016) Effects of the variation of ferroelectric properties on negative capacitance FET characteristics. *IEEE Transactions on Electron Devices* 63(5): 2197–2199.

34. Li, J., et al. (2018) Negative capacitance Ge PFETs for performance improvement: Impact of thickness of HfZrOx. *IEEE Transactions on Electron Devices* 65(3): 1217–1222.

35. Yeung, C.W., Khan, A.I., Sarkar, A. Salahuddin, S., Hu, C. (2013) Low power negative capacitance FETs for future quantum-well body technology. In: *2013 International Symposium on VLSI Technology, Systems and Application (VLSI-TSA)*, pp. 1–2.

36. Appleby, D., Ponon, N.K., Kwa, K.S.K., Zou, B., Petrov, P.K., Wang, T., Alford, N.M., Neill, A.O. (2014) Experimental observation of negative capacitance in ferroelectrics at room temperature. *Nanoscale Letters* 14(7): 3864–3868.

37. Khan, A.I., Radhakrishna, U., Salahuddin, S., Antoniadis, D. (2017) Work function engineering for performance improvement in leaky negative capacitance FETs. *IEEE Electron Device Letters* 38(9): 1335–1338.

38. Ilatikhameneh, H., Ameen, T.A., Chen, C., Klimeck, G., Rahman, R. (2018) Sensitivity challenge of steep transistors. IEEE Transactions of Electron Devices 65(4): 1633–1639.

39. Khan, A.I., Yeung, C.W., Hu, C., Salahuddin, S. (2011) Ferroelectricnegative capacitance MOSFET: Capacitance tuning & antiferroelectric operation. In: *Electron Devices Meeting (IEDM), 2011 IEEE International*, IEEE.

40. Salahuddin, S., Datta, S. (2008) Use of negative capacitance to provide voltage amplification for low power nanoscale devices. *Nanoscale Letters* 8(2): 405–410.

41. Lee, H., Yoon, Y., Shin, C. (2017) Current-voltage model for negative capacitance field-effect transistors. *IEEE Electron Device Letters* 38(5): 669–672.

42. Taur, Y., Ning, T.H. (1998) *Fundamentals of Modern VLSI Devices*. Cambridge University Press, Cambridge, UK.

43. Rahi, S.B., Tayal, S., Upadhyay, A.K. (2021) A review on emerging negative capacitance field effect transistor for low power electronics. *Microelectronics Journal* 116: 105242.

44. Hong, Y., Choi, Y., Shin, C. (2020) NCFET-based 6-T SRAM: Yield estimation based on variation-aware sensitivity. *IEEE Journal of the Electron Devices Society* 8: 182–188.

45. Rahi, S.B., Asthana, P., Gupta, S. (2017) Hetero gate junction less tunnel field-effect transistor: future of low-power devices. *Journal of Computational Electronics* 16(1): 30–38.

46. Upadhyay, A.K., Rahi, S.B., Tayal, S., Song, Y.S. (2022) Recent progress on negative capacitance tunnel FET for low-power applications: Device perspective. *Microelectronics Journal* 129: 105583.

47. Tayal, S., Smaani, B., Rahi, S.B., Upadhyay, A.K., Bhattacharya, S., Ajayan, J., Jena, B., Park, B.-G., Song, Y.S. (2022) Incorporating bottom-up approach into device/circuit co-design for SRAM based cache memory applications. *IEEE Transactions on Electron Devices* 69(11): 6127–6132.

48. Young Suh Song, T.S., Samuel, A., Vimala, P., Tayal, S., Dutta, R., Pandey, C.K., Upadhyay, A.K., Rahi, S.B. (2023) TFET-based memory cell design with top-down approach. In: "*Tunneling Field Effect Transistor: Physics, Design, Modeling and Applications*. CRC.

49. Kumar, D., Rahi, S.B., Paras, N. (2022) Performance analysis of tunnel field effect transistors for low power applications. In: *Intelligent Green Technologies for Sustainable Smart Cities (Advances in Cyber Security)*. Wiley.

50. Upadhyay, A.K., Rahi, S .B., Tayal, S., Song, Y.S. (2022) Recent progress on negative capacitance tunnel FET for low-power applications: Device perspective. *Microelectronics Journal* 129: 105583.

51. Chandrakasan, A., Brodersen, R.W. (1998) Portable terminal electronics. In: *Low-Power CMOS Design*, pp. 367–367. IEEE.

52. Omura, Y. Mallik, A., Matsuo, N. (2016) History of low-voltage and low-power devices. In: *MOS Devices for Low-Voltage and Low-Energy Applications*, pp. 5–11. IEEE.

53. Kobayashi, M., Jin, C. , Hiramoto, T. (2019) Comprehensive understanding of negative capacitance FET from the perspective of transient ferroelectric model. In: *2019 IEEE 13th International Conference on ASIC (ASICON)*, pp. 1–4.

54. Li, Y., Kang, Y., Gong, X. (2017) Evaluation of negative capacitance ferroelectric MOSFET for analog circuit applications. *IEEE Transactions on Electron Devices* 64(10): 4317–4321.

3 Negative Capacitance Field Effect Transistors
Concept and Technology

Ball Mukund Mani Tripathi[1]
[1]Velagapudi Ramakrishna Siddhartha Engineering College, Vijayawada, Andhra Pradesh, India, 520007

3.1 INTRODUCTION

CMOS technology is ubiquitous and plays an important part in our daily lives. Mobile technology is one of the best examples of its application. Scaling has been the main and most important feature of CMOS devices. This feature of CMOS devices has continuously helped in the development of various types of circuits and systems for our daily uses, and also in the medical sciences, aerospace, and military areas for more than four decades. The scaling of various conventional MOSFET types of circuits and systems has been developed to make quality of life improvements [1–5]. The scaling of MOSFET technology is within the 5 nm range. As conventional MOSFET devices are scaled for various uses in VLSI circuits and systems, the off-state current (I_{off}) increases exponentially and is an important aspect for low-power technology. This increase in the power consumption of CMOS circuits and systems [1–20] is illustrated in Figure 3.1. The high off-state current in a conventional MOSFET devices is one of its main limitations. Another limitation of conventional CMOS technology is the minimum 60-mV/decade subthreshold swing as

measured by $SS = \left(\dfrac{d\left(\log_{10} I_{DS}\right)}{dV_{GS}} \right)^{-1} = 2.3 \dfrac{kT}{q} \left(1 + \dfrac{C_D}{C_{OX}} \right)$, which fundamentally limits

the reduction of off-state current [5–15] as well as limiting the scaling of the power supply (V_{DD}). Here, these variables are used for measurement, and are also useful for technical analysis. Thanks to the invention and experimental observation of negative capacitance (NC) in ferroelectric material, various structural and material developments have been implemented [20–25]. It provides practical device design guidelines for low-voltage operation of steep-swing at sub-0.6 V supply voltage. In ultra-scaled conventional MOS devices, the subthreshold swing (SS) reaches the fundamental thermal limit of 60 mV/decade at room temperature, as shown in Figure 3.2. This limitation is an obstacle to achieving high-performance and low-power consumption devices. To solve this issue, various new device concepts have been identified, such as impact ionization MOS (I-MOS) and tunneling FETs (TFETs), which have been suggested due to having a low subthreshold swing.

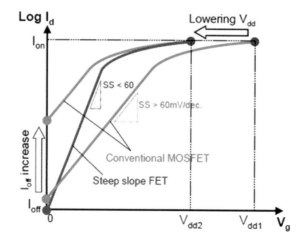

FIGURE 3.1 Limitations of conventional CMOS technology.

Despite its outstanding features, I-MOS suffers from reliability issues and is not suitable for low-power consumption devices because of its high applied voltages [1–30]. In the case of TFETs, on-state current (I_{ON}) is restricted due to band-to-band tunneling probability, even though sub-60-mV/decade switching behavior occurs [3,4,7,8], as shown in Figure 3.3

One important measurement of the efficiency of a MOSFET is the subthreshold swing (SS), which is the change in gate voltage needed to increase the drain current by one order of magnitude. This is measured in units of millivolts per decade in conventional MOSFETs. It is limited to $k_B T/q$ by the Boltzmann electron energy distribution. The numeric value in the ideal case for CMOS devices is ~60 mV/decade at room temperature (T = 300K). In ultra-scaled devices, this limit has become more onerous.

The research chart in Figure 3.2 indicates that there are two optimization domains for the limitation of conventional CMOS devices. The first domain is the transport factor denoted by the letter 'n.' In this research sector, experts examine the transport mechanism of the device and identify ideas for developing desirable results. Strain technology of CMOS has been used for improving the transport of electrons. However, in the case of low power, the tunnel FET is the most popular invention. Here, by using band-to-band tunneling, the device advantages are optimized. The subthreshold swing and I_{OFF} have been found to be at a desirable level, but I_{ON} is limited due to band-to-band transport of charge and a limited tunneling window. Another domain of optimization of CMOS technology limitations for ultra-low-power uses is the body factor, denoted by the letter 'm.' In this research domain, the channel electric field of CMOS devices has been considered for optimization. The invention and development of negative capacitance-based devices and technology are considered in this domain. Bulk NC FET and NC-FinFET are example within this category. Some experts use a hybrid mode of optimization and research for ultra-low-power applications. In this category, transport and body factors are taken into consideration. The NC-tunnel FET is the best example in this context. Experts

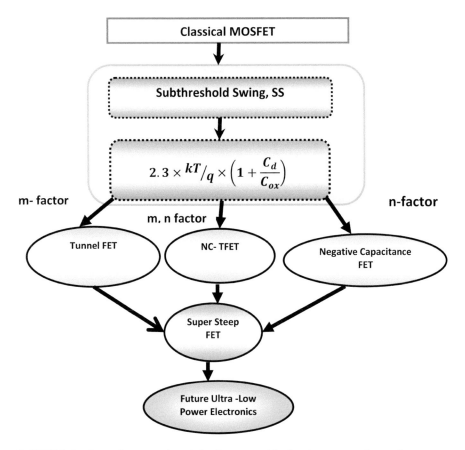

FIGURE 3.2   Scope for research and development with ultra-low-power electronics.

are also using more advanced devices including negative NC-GAA and nanosheet FETs [15–25].

To overcome the challenges posed by conventional CMOS technology for ultra-low-power uses, semiconductor experts have invented tunnel FETs with band-to-band tunneling of charge carriers. This transport mechanism surpasses the subthreshold swing limitation of 60 mV/decade, allowing a lower supply voltage. However, the low on-state current limits its universal adoptability for ultra-low-power uses. These limitations have resulted in the search for alternatives [3–12]. Identification and experimental verification of negative capacitance in ferroelectric materials have been the best solution to overcome the limitations of classical MOS devices and fulfill the modern requirements of ultra-low-power VLSI applications. In ferroelectric-based devices, a ferroelectric capacitor is placed in series with a conventional dielectric gate capacitor [20–38]. For example, in a ferroelectric tunnel (FET) an ultra-steep (abrupt) switch has a better subthreshold swing (SS) than the MOSFET limit of 60 mV/decade at room temperature. This device combines two key principles: ferroelectric gate stack and band-to-band tunneling in gated p-i-n junction, in which the ferroelectric

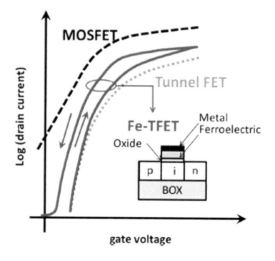

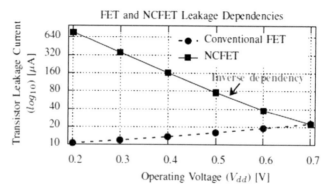

FIGURE 3.3   Qualitative analysis of conventional MOSFET with tunnel FET and impact of ferroelectric on the device, and other comparisons.

material included in the gate stack creates, due to dipole polarization with increasing gate voltage, a positive feedback in the capacitive coupling that controls the band-to-band (BTBT) tunneling at the source junction of the silicon p-i-n reversed bias structure, where the combined effect of band-to-band tunneling and ferroelectric negative capacitance offers more abrupt off–on and on–off transitions, as shown in Figure 3.3.

In the remainder of this chapter, the negative capacitance-based developments are discussed.

3.2   NEGATIVE CAPACITANCE TECHNOLOGY

Salahuddin and Datta's group identified negative capacitance in ferroelectric materials. This gained significant attention in the low-power research world. The negative capacitance effect reduces the required supply voltage, known as the Boltzmann limit, in conventional complementary metal–oxide–semiconductor transistors by amplifying the internal voltage by reducing the internal capacitance of MOSFETs [1–12]. The thermodynamic perspective to understand how the reduction in energy dissipation is achieved and the scope of low-power development with NC FET technology is shown in Figure 3.3. As illustrated in Figure 3.3, the negative capacitance field effect transistor (NCFET) is one of the most promising emerging technologies to overcome the fundamental limits of conventional CMOS technology. Due to the ability of the internal amplification and low leakage current, NCFETs have gained significant interest. The NCFET features a ferroelectric (FE) layer within the transistor's gate, which internally amplifies the voltage, allowing the NCFET to operate at a lower voltage while sustaining performance with considerable energy savings. The NCFET technology, with a thick ferroelectric layer (FE), has a voltage reduction which increases the leakage power, rather than decreasing it, due to the negative drain-induced barrier lowering (DIBL) effect. The NCFET is an emerging technology that incorporates a ferroelectric layer within the transistor gate stack to overcome the fundamental limit of sub-threshold swing in transistors [16–30].

In the sub-threshold region, the value of the capacitance of ferroelectric layer (C_{FE}) should be as close as possible to (but larger than) the MOS oxide capacitance (C_{MOS}) to achieve large potential amplification. However, when strong inversion of the MOS occurs and C_{MOS} increases rapidly, $-C_{FE} > C_{MOS}$ is still required to stabilize the NCFET. The desirable properties can be achieved by utilizing the polysilicon capacitance (C_{poly}), which is a function of the applied voltage. The magnitude of the effective C_{FE}' (=$1/(1/C_{FE} + 1/C_{poly})$) approaches C_{MOS} capacitance. Its value can approach the subthreshold region [20–30]. To maintain the constant gate control it is important to ensure accurate MOSFET adjustment of the drive strength by changing its size. In the NCFET, the non-uniform distribution of ferroelectric polarization and capacitance match are sensitive to the size and tend to increase with the fluctuation of gate control [30–45].

The idea of using a ferroelectric (FE) material for transistors originates from the realization that internal dipolar interactions in an FE material can be exploited to provide an amplification of the electric field. In ferroelectric materials with the same energy difference between two states, N (number of electric dipoles being switched) and lower energy are required compared to a typical dielectric for switching. Thus, it can be understood that materials with interacting variables can switch their order with less energy compared to non-interacting state variables. Using this concept, the energy dissipation can be reduced by N times, where N is the number of charges that are 'switched' in each conventional transistor. In addition, the concept can be used to reduce the voltage needed for switching transistors between the ON and OFF states, and thus reducing energy dissipation in transistor switching. The two polarization states and corresponding energy affect the polarization of a ferroelectric material. The two polarization vectors are separated by an energy barrier. The energy function can be written as shown in Equation (3.1), where U represents the free energy of the ferroelectric material and is defined by the Landau–Devonshire theory:

$$U = \alpha P^2 + \beta P^4 + \gamma P^6 \qquad (3.1)$$

In Equation (3.1), P is the polarization and α, β, and γ are the L-K constants. The charge across the ferroelectric material is shown in Equation (3.2), where E is an

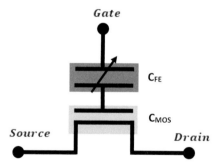

FIGURE 3.4 Equivalent capacitance model of NCFET.

external electric field and α, β, and γ are the order parameters. In the case of the ferro-electric material, α is always negative.

$$Q_{FE} = \varepsilon_0 E_{FE} + P \qquad (3.2)$$

In Equa tion (3.2), ε_0 is the vacuum permittivity, and E_{FE} is the electric field across the ferroelectric material. In the conventional MOSFET, when the ferroelec-tric material is used as an insulator material on top of an Si channel, the overall electrostatic picture can be described as a series combination of two capacitors: (1) the ferroelectric capacitor denoted as C_{FE} and (2) the capacitor C_{MOS} represents the channel-to-drain capacitance, with other fringe capacitances between the channel and ground as shown in Figure 3.4. It is assumed that the ferroelectric is in steady-state polarization ($dP/dt = 0$).

These two series capacitors are nonlinear unlike the insulator capacitance in a standard MOSFET which is a simple linear dielectric capacitance having a parabolic energy equation (i.e., $Q^2/2C$ where Q is the charge and C is the capacitance). Since both the capacitances are in series, the total charge in both has to be the same: $Q_{FE} = Q_{surface}$ where Q_{FE} and $Q_{surface}$ are the charges of the C_{FE} and C_{MOS} capacitors, respectively. Figure 3.5 shows a conventional schematic diagram for an NCFET device and represents the measurement of negative capacitance in ferroelectric materials.

The total voltage shared by the two capacitors is equal to the gate voltage summation. In Equation (3.3), V_{FE} and $\psi_{surface}$ are the voltages across the FE and the semiconductor capacitor, respectively.

$$Vg = V_{FE} + \psi_{surface} \qquad (3.3)$$

The I–V curve is steeper in the subthreshold region and goes beyond the Boltzmann limit at room temperature, compared to a conventional device without ferroelectric material (MOSFET). This indicates that the same current can be obtained at a lower gate voltage and therefore with less energy dissipation for a given speed of operation compared to a conventional MOSFET.

Figure 3.6 shows the impact on negative capacitance in transfer characteristics of NCFET. From Figure 3.6, it can be observed that, at the same operating voltage,

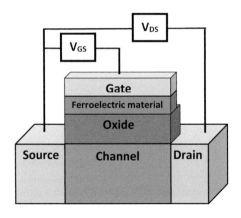

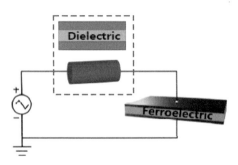

FIGURE 3.5 Schematic of conventional NCFET and measurement of negative capacitance in ferroelectric materials.

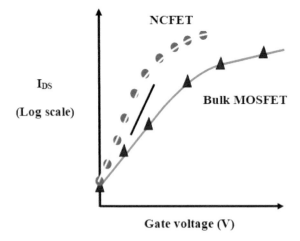

FIGURE 3.6 The I–V characteristic of a Si MOSFET with and without FE capacitance.

NCFET shows lower SS than conventional MOSFET with similar current. From this figure, it can be clearly observed that ferroelectric material in NCFET reduces the threshold voltage of devices with a similar off-state current. A lower threshold voltage with similar off-current indicates a low operational behavior, which is possible only with voltage amplification.

3.3 PSEUDO-ANALYTICAL SOLUTION PROCEDURE FOR NEGATIVE CAPACITANCE

In addition to a theoretical point of view, it often provides a better understanding if an analytical solution procedure is provided for a phenomenon. The pseudo-analytical solution gives a deeper insight into the effects of negative capacitance, such as steeper I–V characteristics. In NCFET having conventional MOSFET as a base device, as shown in Figure 3.5, the surface potential can be estimated by Equation (3.4).

$$\psi_{surface} = V_g - V_{FE} \tag{3.4}$$

In the device, $C_{surface}$ is a strong function of voltage and for simplicity is given as $Q_{surface}/\psi_{surface}$, so that the charge can be written as shown in Equation (3.5):

$$Q_{surface} = C_{surface} \times \psi_{surface} = C_{surface} \times (V_g - V_{FE}) \tag{3.5}$$

By using the charge conservation principle, the series capacitors must have the same charge:

$$Q_{FE} = \varepsilon_0 E_{FE} + P = C_{surface}(V_g - V_{FE}) = Q_{surface} \tag{3.6}$$

Therefore, a load line can be drawn as:

$$P = C_{surface}[V_g - (1 + \frac{ts}{\mu r\, tFE})V_{FE}] \tag{3.7}$$

In Equation (3.7), ε_r is the relative permittivity of dielectric $C_{surface}$ and thickness, and t_{FE} is the thickness of the ferroelectric layer. The plot of P versus V_{FE} is the hysteresis curve and resembles an S curve. The right-hand side acts as a load line on the S curve. The slope of the load line shown in the figure is $C_{surface}[1 + t_s/(\_rtFE)]$, and the intersection with the horizontal axis is $V_g[1 + ts/(\varepsilon_r t_{FE})]^{-1}$. Therefore, if the load line intersects the negative slope region of the S curve, it gives a self-consistent solution for the nonlinear electrostatics problem of the two series capacitors, where the FE itself acts in the region where $\partial P/\partial E_{FE}$ is negative. The capacitance of the FE layer is given as Equation (3.8):

$$C_{FE} = \frac{\partial Q_{FE}}{\partial V_{FE}} = \frac{\varepsilon o}{t_{FE}}(1 + \frac{1}{\varepsilon_0}\frac{\partial p}{\partial E_{FE}}) \tag{3.8}$$

Therefore, the ferroelectric layer will show a negative capacitance effect if it follows the condition shown in Equation (3.9)

$$\left| \frac{1}{\varepsilon_0} \frac{\partial p}{\partial E_{FE}} \right| >> 1 \tag{3.9}$$

Now, let's consider the following equation:

$$C_{surface.} \left(1 + \frac{1}{s_r} \frac{t_s}{t_{FE}} \right) > \left| \frac{\partial P}{\partial E_{FE}} \right| \tag{3.10}$$

When Equation (3.10) is satisfied, the load line intersects the S curve at multiple points, which denotes a hysteretic behavior. This leads to negative capacitance for the ferroelectric layer. The rate of change in the surface potential with respect to the gate voltage is given in Equation (3.11)

$$\frac{\partial \psi_{Surface}}{\partial V_g} = \frac{|C_{FE}|}{|C_{FE}| - C_{surface}} > 1 \tag{3.11}$$

Equation (3.11) shows that the change in the surface potential is amplified compared to the supply voltage, unlike in a conventional series combination of two capacitors where the change in the internal node voltage will always be smaller than the supply voltage. This enhanced surface potential because of the negative capacitance is responsible for steeper I–V characteristics in MOSFETs. Also, hysteretic behavior occurs when $|C_{FE}| < C_{surface}$ and $\frac{\partial \psi_{Surface}}{\partial V_g} < 0$. In addition, it should be noted that even if $\frac{\partial p}{\partial V_{FE}} < 0$, the ferroelectric polarization does not change its direction on applying an external voltage. In conclusion, the polarization always changes in the same direction as that of the applied field. However, the resultant field, which is summation of the internal and external fields across the ferroelectric, is negative in this region. Generally, negative capacitance transistors are analyzed using the ferroelectricity model. The mathematical series resulting from the expansion of Landau provides the information about the transformations in the crystal symmetry. Every point in the S curve corresponds to microscopic configurations of relative ion displacements. Integrating the electric field with respect to polarization, the potential energy landscape is obtained, resembling the standard Landau free energy landscape at zero temperature [31–55].

3.4   NEGATIVE CAPACITANCE STABILITY

As the total gate capacitance is the sum of oxide and FE capacitances, the oxide capacitance is assumed to be a linear dielectric capacitance, as denoted by Equation (3.12):

$$C_{surface} = Q_{surface}/\psi_{surface} \tag{3.12}$$

Also, charge across the ferroelectric materials can be estimated by Equation (3.13)

$$Q_{FE} = \varepsilon_0 E_{FE} + P = \varepsilon_0 \varepsilon_r E_{surface} = Q_{surface} \tag{3.13}$$

At zero gate voltage, the dielectric field $E_{surface}$ and FE fields are related as

$$E_{surface} = -(t_{FE}/t_s)\, E_{FE} \tag{3.14}$$

Therefore,

$$E_{FE} = \frac{p/\epsilon_0}{1 + \dfrac{\epsilon_0\, t_{FE}}{t_s}} \tag{3.15}$$

Again, since,

$$E_{FE} = \left(Q_{FE} - P\right)/\varepsilon_0 \tag{3.16}$$

and

$$Q = Q_{FE} = Q_{surface} \tag{3.17}$$

the total charge on plates Q can be expressed as:

$$Q = \frac{p}{1 + \dfrac{t_s}{t_{FE}\, \epsilon r}} \tag{3.18}$$

This equation shows that the charge is a function of the nature of the dielectric as well as its thickness, and decreases with increasing dielectric thickness. The reduction in effective charge with introduction of a dielectric layer induces depolarization of the field. The addition of the depolarizing field results in changing the net force toward zero at smaller polarizations. Effectively, it can be perceived as the effect of the depolarization field , suppressing polarization. The effect of the depolarization field is common in ferroelectric memory devices and methods have been suggested to reduce this effect such as by increasing the thickness of the FE layer which reduces the field. Similarly, the formation of an unintended dead layer of dielectric between the FE and metal electrode will reduce this field. The use of oxide electrode prevents the formation of a dead layer. It has been seen that when the depolarization field is very strong the only available solution where the net force is zero is where the polarization is also zero. This is the region for negative capacitance operation. It has been observed that by applying an external field on the unperturbed force–displacement curve to increase the polarization from this point, the FE will operate in the region where the net polarizing field is larger than the net depolarizing field inside the ferroelectric layer. This enhanced polarization boosts the electric field at the FE–dielectric (FE–DE) interface. Also, the presence of two stable points where the net force can be zero shows that there will be hysteresis. However, the polarization is still not high enough and the ferroelectric is stable. In this region, the ferroelectric shows negative capacitance with low hysteresis. Although it is not ideal for digital logic applications it is worth analyzing it experimentally. The strength of the depolarization field needs to stabilize the ferroelectric at zero polarization [31–55].

It has been observed that the ferroelectric becomes stable in a region of the configuration space in which the net force on an average dipole is zero. In addition, every state in the negative capacitance region corresponds with the physically defined structures of the unit cell. In summary, it can be said that it is better if the ferroelectric smoothly traverses the configuration space to transition to a zero-polarization state and, for this, a second-order phase transition material is better than a first-order phase transition material. Also, it has been observed that, in the presence of substrate, strain is present, and the ferroelectric transition to zero polarization is difficult as it acts against the substrate strain and therefore, epitaxially grown hetero structures are preferable for the ferroelectric to break into domains and stabilize the negative capacitance state in the domain walls. However, for poly-crystalline ferroelectrics where such constraints from the substrate are not present, it is easier for the entire volume of the material to stabilize in the negative capacitance state. Therefore, it can be seen that polycrystalline FEs such as doped HfO_2 are more suitable for negative capacitance operation. It is better to sandwich the ferroelectric layer with an amorphous dielectric layer, which easily accommodates changes in strain.

It has been observed that below a certain length of the polarization vector, the ferroelectric operates in a region where the force due to the dipolar field dominates over the mechanical contribution. The field across the ferroelectric is opposite to external field and the overall $1/C = dV/dQ$ is negative. Although the analysis has been done for a single unit cell and considering an average effect of the surrounding unit cells, the spatial distribution of the polarization is not uniform or like in a single domain. In fact, the ferroelectric film breaks down into domains depending on the electrostatics of the system. However, if the polarization vectors are below the threshold length for negative capacitance, regions of local negative capacitance arise. Therefore, in FE-DE heterostructures, considering the FE polarization is completely uniform, only the capacitance of the FE region is negative and the total capacitance is always positive. Even if the FE is broken into domains, the local regions of negative and positive capacitances could arise in the FE film itself and in conjunction with a series dielectric this provides greater electrostatic potential across the dielectric.

The state of negative capacitance is stabilized because the series combination of a ferroelectric and dielectric indicates a non-zero depolarization field in the FE layer. At zero gate voltage, the surface voltage and voltage across the FE may not be zero, but the sum must be zero according to Kirchhoff's law. However, in the presence of a significant leakage through the dielectric and FE layers, the voltage across these layers will be zero due to a short circuit. Thus, there is no depolarization field, and therefore, no negative capacitance stabilization. In a MFIS (metal–FE-metal–insulator–semiconductor) type of structure, the FE layer is short circuited due to the internal metal layer in the presence of a leakage, therefore no dc stabilization of negative capacitance is possible at a zero gate voltage. The FE shorting occurs due to the presence of significant charge density at the FE–DE interface originating from traps or tunneling of electrons. When the gate voltage is non-zero dc stabilization is still possible. Precautions are desired while studying negative capacitance experimentally, particularly when high voltages are applied across an MFIS structure with a very thin oxide (insulator) layer because the high electric field across the structure may melt down or high leakage may occur. Therefore, it can be suggested that capacitance amplification and transistor studies should be done at a voltage much lower than the breakdown voltage of the dielectric used in the structure for experimental study.

3.5   FERROELECTRIC MATERIALS AND THEIR COMPATIBILITY WITH NEGATIVE CAPACITANCE TECHNOLOGY

It has been observed that the shape of the S curve varies according to the atomic coordinates of the dipoles and the local potential energy of the lattice. Therefore, providing a quantitative model which is necessary to predict the device design is not an easy task. Generally, we take a $P–V$ curve of a thin-film FE, fit a Landau free energy model to it, and then construct the S curve. This approach is good for a qualitative analysis but not for a quantitative one, especially for the information when the slope of the S curve changes from negative to positive. In common practice, a set of experiments with varying thickness combinations in FE–DE hetero structures are done, and then the results are compared with the dielectric capacitance, to obtain the S curve. Generally, ferroelectric materials for NC transistors are required to fulfill the following conditions to be useful for a real device:

- The material should be able to show robust ferroelectricity below 5 nm thickness.
- It should be compatible with CMOS technology.
- The layer of FE material should be thermally stable on silicon.
- The deposition should be conformal on 3D substrates.
- It should have a large bandgap energy and conduction band offset to Si.

A lower ferroelectric thickness helps to stabilize the NC state and reduces the distance between neighboring NCFETs in FinFET technology. The thickness is even more crucial for nanowire or nanosheet devices. In addition, these materials must be compatible with the silicon CMOS process flow. In the CMOS process, the ferroelectric materials should be able to withstand high temperatures as well as hydrogen annealing. Furthermore, there should be mature disposition techniques used that give conformal and homogeneous deposition for ultra-thin ferroelectric films for large wafers. For minimizing leakage, a high electronic bandgap as well as high conduction band offset to Si are also required.

3.6   TRANSISTORS WITH NEGATIVE CAPACITANCE

To understand the complex interaction between the FE material in the gate stack and the underlying FET (baseline FET) capacitance, device level modeling of FeFETs can provide very important insights. In this regard, transistor equations (Poisson's equation and the drift-diffusion current continuity equations) must be solved self-consistently with the Landau-Khalatnikov (L-K) equation of the FE. After the discovery of doped HfO_2 as a ferroelectric oxide, there was the possibility to integrate an FE oxide in the transistor in the gate-stack for commercial production. Recently, a many transistors with negative capacitance have been demonstrated. The same concept was demonstrated for the first time with Si FinFET transistors with a smaller geometry. Although a lot of research and development has done, the critical area for the realization of negative capacitance in transistors is related to capacitance patching, current amplification, and the various FE materials used. Capacitance matching is important for a MOSFET structure because the dielectric capacitor in the FE-DE hetero structure is related to stabilization in the negative capacitance state [31–55]. The semiconductor capacitance increases rapidly from a small value at low voltage to a higher value at larger gate voltages. This capacitance $C_{surface}$ changes by two orders of magnitude from OFF to ON in MOSFET, which is contradictory to the requirement of a negative capacitance transistor design. For a lower subthreshold swing than 60 mV/decade, the $|C_{FE}|$ should be matched with the $C_{surface}$ at a low gate voltage. This implies that when the transistor turns ON, $|C_{FE}|<C_{surface}$, there is a hysteretic region of operation, which is highly undesirable [20–35].

3.7   FERROELECTRICS IN THE GATE STACK OF A TRANSISTOR

NCFETs are one of the most promising emerging technologies that may overcome the fundamental limits of conventional CMOS technology. Due to the internal voltage amplification, NCFETs have drawn significant interest. NCFETs feature a ferroelectric (FE) layer within the transistor's gate, which internally amplifies the voltage, allowing the NCFET to operate at a lower voltage while sustaining performance with resulting considerable energy savings. In NCFET technology with a thick ferroelectric layer, voltage reduction increases the leakage power, rather than decreasing it, due to the negative drain-induced barrier lowering (DIBL) effect [1–20].

3.8   EXPERIMENTAL INVESTIGATION OF NEGATIVE CAPACITANCE

The total capacitance of the ferroelectric layer which is stabilized in a negative capacitance region and the series connected dielectric capacitance is greater than the dielectric capacitance itself and is expressed by Equation (3.19):

$$\frac{1}{C_t} = \frac{1}{C_{FE}} - \frac{1}{C_{DI}} \qquad (3.19)$$

Therefore,

$$C_{\mathrm{T}} > C_{\mathrm{DI}} \text{ if } C_{FE} > C_{DI}$$

In Equation (3.19), C_{T}, C_{FE}, and C_{DI} are total, FE, and dielectric capacitance, respectively. This equation shows that different matching combinations can be created by changing the FE capacitance. The one matching combination is when $|C_{FE}| < C_{DI}$, which results in capacitance enhancement with hysteresis. In the second case, when $|C_{FE}| > C_{DI}$, the capacitance will be enhanced with no or minimal hysteresis. Thus, by carefully tuning the value of C_{FE}, the magnitude of capacitance amplification as well as hysteresis can be change. The first demonstration of capacitance amplification using an epitaxially grown bilayer FE was PZT. It is evident that the FE properties of the single-crystalline PZT can be controlled by changing the temperature, thus resulting in different capacitance matching conditions. Capacitance amplification has been demonstrated using different FEs. It also has been demonstrated that the contribution of domain wall polarization is most important for negative capacitance because it is more favorable energetically for these dipoles to get suppressed. Recently, many other groups have shown capacitance enhancement in FE-DE hetero structures.

With the progress in silicon circuit miniaturization, lowering power consumption becomes the major objective. Supply voltage scaling in ultra large-scale integration (ULSI) is limited by the physical barrier termed 'Boltzmann tyranny.' Moreover, considerable heat is inevitably generated from the ultra-highly integrated circuit. To solve these problems, a ferroelectric negative capacitance field effect transistor (Fe-NCFET) is proposed in order to reduce the subthreshold swing (SS) through an internal voltage amplification mechanism, thus effectively scaling the supply voltage and significantly lowering the power dissipation of ULSI. In this review, representative research results on NCFETs are comprehensively reviewed to suggest benefits for further study. Here, the background and significance of NCFETs are introduced, and the physical essence of the negative capacitance effect is reviewed. Then, physical models and simulation methods of NCFETs are classified and discussed under the consideration of three basic gate structures. Several influencing factors of device performance—SS, on–off ratio, and hysteresis—are also theoretically analyzed. Moreover, the experimental results of NCFETs based on different ferroelectric materials are summarized. Finally, with the combination of NC effect and two-dimensional materials (FinFET) and tunneling FET, respectively, several novel and potential NCFETs are presented, and the outlook for NCFETs is proposed.

3.9   TEMPERATURE EFFECT

The impact of temperature variation on negative capacitance FET is an important issue. It has been observed experimentally that the internal voltage amplification peaks of MFMIS structures depend on variation of the S-shape of the polarization with respect to electrical field characteristics. This amplification of internal voltage causes a reduction in the subthreshold swing in MFMIS structures. However, this effect is ineffective at a temperature near to the Curie temperature because of the shrinking of the NC region and saturation of the amplification. It is also reported that counter-clockwise rotation of the P-V loops increases the *dP/dV* slope with the temperature. This is equivalent to an increase in the overall ferroelectric capacitance with temperature. In conclusion, it has been experimentally and theoretically demonstrated that the maximum amplification occurs at an optimum temperature.

3.10   CONCLUSION

The negative capacitance field effect transistor (NCFET) is a promising technology which exhibits lower subthreshold swing (SS) and high ON current beyond the limit of conventional CMOS. The key feature of NCFETs are its sub-threshold slope (SS) <60 mV/decade at 300 K. NCFETs are a leading emerging technology that promises outstanding performance in addition to better energy efficiency. The thickness of the added ferroelectric layer as well as the frequency and voltage are the key parameters that impact the power and energy of NCFET-based processors in addition to the characteristics of runtime workloads. NCFETs boost the electric field at the semiconductor–channel interface by virtue of the gate voltage amplification effect of a ferroelectric (FE) layer.

REFERENCES

1. R. C. Bheemana, A. Japa, S. Yellampalli and R. Vaddi (2021) Steep switching NCFET based logic for future energy efficient electronics. *2021 IEEE International Symposium on Smart Electronic Systems (iSES)*, pp. 327–330.
2. Y. Hong and C. Shin (2020) Yield estimation of NCFET-based 6-T SRAM. *2020 4th IEEE Electron Devices Technology & Manufacturing Conference (EDTM)*, pp. 1–3.
3. S. B. Rahi, S. Tayal, A. Kumar (2021) Emerging negative capacitance field effect transistor in low power electronics. *Microelectronics Journal* 116: 105242.
4. A. K. Upadhyay, S. B. Rahi, S. Tayal, Y. S. Song (2022) Recent progress on negative capacitance tunnel FET for low-power applications: device perspective. *Microelectronics Journal* 129: 105583.
5. S. Salamin, M. Rapp, J. Henkel, A. Gerstlauer and H. Amrouch (2022) Dynamic power and energy management for NCFET-based processors. *IEEE Transactions on Computer-Aided Design of Integrated Circuits and Systems* 39(11): 3361–3372.
6. Y. Hong, Y. Choi and C. Shin (2020) NCFET-based 6-T SRAM: yield estimation based on variation-aware sensitivity. *IEEE Journal of the Electron Devices Society* 8: 182–188.
7. S. Tayal, S.B. Rahi, J. P. Srivastava and S. Bhattacharya (2021) Recent trends in compact modeling of negative capacitance field effect transistors. In: *Semiconductor Devices and Technologies for Future Ultra-low Power Electronics*. CRC
8. Tayal, Shubham, Abhishek Kumar Upadhyay, Deepak Kumar, and Shiromani Balmukund Rahi (eds.) (2022) *Emerging Low-Power Semiconductor Devices: Applications for Future Technology Nodes*. CRC Press.

9. K. Lee et al. (2019) Analysis on fully depleted negative capacitance field-effect transistor (NCFET) based on electrostatic potential difference. In: *2019 Electron Devices Technology and Manufacturing Conference (EDTM)*, pp. 422–424.

10. X. Sun, Y. Zhang, J. Xiang, K. Han, X. Wang and W. Wang (2021) Role of interfacial traps at SiO_2/Si interface in negative capacitance field effect transistor (NCFET) based on transient negative capacitance (NC) theory. In: *2021 5th IEEE Electron Devices Technology & Manufacturing Conference (EDTM)*, pp. 1–3.

11. S. Salamin et al. (2021) Power-efficient heterogeneous many-core design with NCFET technology. *IEEE Transactions on Computers* 70(9): 1484–1497.

12. S. Roy, P. Chakrabarty and R. Paily (2022) Assessing RF/AC performance and linearity analysis of NCFET in CMOS-compatible thin-body FDSOI. *IEEE Transactions on Electron Devices* 69(2): 475–481.

13. M. Gu et al. (2021) FDSOI NCFET with stepped thickness ferroelectric layer. In: *2021 5th IEEE Electron Devices Technology & Manufacturing Conference (EDTM)*, pp. 1–3.

14. H. Wang et al. 2018) New insights into the physical origin of negative capacitance and hysteresis in NCFETs. In: *2018 IEEE International Electron Devices Meeting (IEDM)*, pp. 31.1.1–31.1.4.

15. Z. C. Yuan et al. (2019) Toward microwave S- and X-parameter approaches for the characterization of ferroelectrics for applications in FeFETs and NCFETs. *IEEE Transactions on Electron Devices* 66(4): 2028–2035.

16. B. Obradovic, T. Rakshit, R. Hatcher, J. A. Kittl and M. S. Rodder (2018) Ferroelectric switching delay as cause of negative capacitance and the implications to NCFETs. In: *2018 IEEE Symposium on VLSI Technology*, pp. 51–52.

17. C.-C. Fan, C.-H. Cheng, Y.-R. Chen, C. Liu and C.-Y. Chang (2017) Energy-efficient HfAlOx NCFET: Using gate strain and defect passivation to realize nearly hysteresis-free sub-25mV/dec switch with ultralow leakage. In: *2017 IEEE International Electron Devices Meeting (IEDM)*, pp. 23.2.1–23.2.4.

18. T. Cam et al. (2020) Sustained benefits of NCFETs under extreme scaling to the end of the IRDS. *IEEE Transactions on Electron Devices* 67(9): 3843–3851.

19. J. K. Wang et al. (2022) Potential enhancement of fT and g_mfT/ID via the use of NCFETs to mitigate the impact of extrinsic parasitics. *IEEE Transactions on Electron Devices* 69(8): 4153–4161.

20. Y. Liang, X. Li, S. K. Gupta, S. Datta and V. Narayanan (2018) Analysis of DIBL effect and negative resistance performance for NCFET based on a compact SPICE model. *IEEE Transactions on Electron Devices* 65(12): 5525–5529.

21. X. Li et al. (2017) Advancing nonvolatile computing with nonvolatile NCFET latches and flip-flops. *IEEE Transactions on Circuits and Systems I: Regular Papers* 64(11): 2907–2919.

22. S. Salamin et al. (2022) Impact of NCFET technology on eliminating the cooling cost and boosting the efficiency of Google TPU. *IEEE Transactions on Computers* 71(4): 906–918.

23. R. A. Vega, T. Ando and T. M. Philip (2021) Junction design and complementary capacitance matching for NCFET CMOS logic. *IEEE Journal of the Electron Devices Society* 9: 691–703.

24. Y. Liang et al. (2018) Influence of body effect on sample-and-hold circuit design using negative capacitance FET. *IEEE Transactions on Electron Devices* 65(9): 3909–3914.

25. Y. Liang et al. (2018) Influence of body effect on sample-and-hold circuit design using negative capacitance FET. *IEEE Transactions on Electron Devices* 65(9): 3909–3914.

26. Prakash, Om, et al. (2021) On the critical role of ferroelectric thickness for negative capacitance device-circuit interaction. *IEEE Journal of the Electron Devices Society* 9: 1262–1268.
27. M. Kim, J. Seo and M. Shin (2018) Biaxial strain based performance modulation of negative-capacitance FETs. In: *2018 International Conference on Simulation of Semiconductor Processes and Devices (SISPAD)*, pp. 318–322.
28. H.-H. Lin and V. P.-H. Hu (2019) Device designs and analog performance analysis for negative-capacitance vertical-tunnel FET. In: *20th International Symposium on Quality Electronic Design (ISQED)*, pp. 241–246.
29. C.-I. Lin, A. I. Khan, S. Salahuddin and C. Hu (2016) Effects of the variation of ferroelectric properties on negative capacitance FET characteristics. *IEEE Transactions on Electron Devices* 63(5): 2197–2199.
30. A. Sharma and K. Roy (2017) Design space exploration of hysteresis-free HfZrOx-based negative capacitance FETs. *IEEE Electron Device Letters* 38(8): 1165–1167.
31. C. Jin, T. Saraya, T. Hiramoto and M. Kobayashi (2019) Transient negative capacitance as cause of reverse drain-induced barrier lowering and negative differential resistance in ferroelectric FETs. In: *2019 Symposium on VLSI Technology*, pp. T220–T221.
32. C.-C. Fan et al. (2018) Interface engineering of ferroelectric negative capacitance FET for hysteresis-free switch and reliability improvement. In: *2018 IEEE International Reliability Physics Symposium (IRPS)*, pp. P-TX.8-1–P-TX.8-5.
33. H. Eslahi, T. J. Hamilton and S. Khandelwal (2021) Circuit performance analysis of analog RF LNA designed with negative capacitance FET. In: *2021 IEEE Asia-Pacific Microwave Conference (APMC)*, pp. 284–286.
34. Luk'yanchuk, I., Razumnaya, A., Sené, A. et al. (2022) The ferroelectric field-effect transistor with negative capacitance. *npj Computational Materials* 8: 52.
35. Cao, W. and Banerjee, K. (2020) Is negative capacitance FET a steep-slope logic switch? *Nature Communications* 11: 196.
36. S. Salahuddin (2016) Review of negative capacitance transistors. In: *2016 International Symposium on VLSI Technology, Systems and Application (VLSI-TSA)*, pp. 1.
37. C. Jiang, L. Zhong and L. Xie (2019) Effects of interface trap charges on the electrical characteristics of back-gated 2D negative capacitance FET. In: *2019 IEEE 19th International Conference on Nanotechnology (IEEE-NANO)*, pp. 163–166.
38. M. Soleimani, N. Asoudegi, P. Khakbaz and M. Pourfath (2019) Negative capacitance field-effect transistor based on a two-dimensional ferroelectric. In: *2019 International Conference on Simulation of Semiconductor Processes and Devices (SISPAD)*, pp. 1–4.
39. D. Kwon et al. (2020) Near threshold capacitance matching in a negative capacitance FET with 1 nm effective oxide thickness gate stack. *IEEE Electron Device Letters* 41(1): 179–182.
40. J. Li et al. (2017) Correlation of gate capacitance with drive current and transconductance in negative capacitance Ge PFETs. *IEEE Electron Device Letters* 38(10): 1500–1503.
41. P. Bidenko, S. Lee, J.-H. Han, J. D. Song and S.-H. Kim (2018) Simulation study on the design of sub-non-hysteretic negative capacitance FET using capacitance matching. *IEEE Journal of the Electron Devices Society* 6: 910–921.
42. P. Bidenko, S. Lee, J.-H. Han, J. D. Song and S.-H. Kim (2018) Simulation study on the design of sub-non-hysteretic negative capacitance FET using capacitance matching. *IEEE Journal of the Electron Devices Society* 6: 910–921.
43. S.-Y. Lee et al. (2020) Effect of seed layer on gate-all-around poly-Si nanowire negative-capacitance FETs with MFMIS and MFIS structures: planar capacitors to 3-D FETs. *IEEE Transactions on Electron Devices* 67(2): 711–716.

44. M. Kobayashi (2020) On the physical mechanism of negative capacitance effect in ferroelectric FET. In: *2020 International Conference on Simulation of Semiconductor Processes and Devices (SISPAD)*, pp. 83–87.

45. J. Zhou et al. (2018) Negative differential resistance in negative capacitance FETs. *IEEE Electron Device Letters* 39(4): 622–625.

46. S.-E. Huang, S.-H. Lin and P. Su (2020) Investigation of inversion charge characteristics and inversion charge loss for InGaAs negative-capacitance double-gate FinFETs considering quantum capacitance. *IEEE Journal of the Electron Devices Society* 8: 105–109.

47. Y. Zhao et al. (2019) Experimental study on the transient response of negative capacitance tunnel FET. In: *2019 Electron Devices Technology and Manufacturing Conference (EDTM)*, pp. 88–90.

48. T. Srimani et al. (2018) Negative capacitance carbon nanotube FETs. *IEEE Electron Device Letters* 39(2): 304–307.

49. S. Dasgupta et al. (2015) Sub-kT/q switching in strong inversion in PbZr0.52Ti0.48O3 gated negative capacitance FETs. *IEEE Journal on Exploratory Solid-State Computational Devices and Circuits* 1: 43–48.

50. S. Guo, R. J. Prentki, K. Jin, C.-l. Chen and H. Guo (2021) Negative-capacitance FET with a cold source. *IEEE Transactions on Electron Devices* 68(2): 911–918.

51. B. M. M. Tripathi, and S. P. Das (2019) Analysis of parameter variations and their impact on hetero-junctionless tunnel field effect transistor due to negative capacitance. *Advanced Science, Engineering and Medicine* 11(8): 728–733.

52. B. Ghosh, et al. (2013) Ultrathin compound semiconductor in bulk planar junctionless transistor for high-performance nanoscale transistors. *Journal of Low Power Electronics* 9(4): 490–495.

53. B. M. M. Tripathi, P. Jain, and S. P. Das. (2017) SiGe source dual metal double gate tunnel field effect transistor. *Journal of Low Power Electronics* 13(1): 76–82.

54. B. Ghosh, et al. (2017) In0 25Ga0 75As channel double gate junctionless transistor. *Journal of Low Power Electronics* 10(1): 101–106.

55. J. C. Wong and S. Salahuddin (2018) Negative capacitance transistors. *Proceedings of the IEEE* 107(1): 49–62.

4 Basic Operation Principle of Negative Capacitance Field Effect Transistor

Malvika[1], Bijit Choudhuri[1],
Kavicharan Mummaneni[1]
[1]Department of Electronics and Communication
Engineering, National Institute of Technology Silchar,
Assam, 788010, India

4.1 INTRODUCTION

4.1.1 BACKGROUND

Over the past 75 years, there has been astounding exponential growth in the efficiency of computations. Since the 1960s, this development has been primarily fueled through the shrinking of metal-oxide semiconductor field effect transistors (MOSFETs), the dimensions (lateral) of which went down to below 100 nanometers in 2003, ushering in the nanoelectronics age. Ever since, new materials and device designs (fin field effect transistors and fully depleted silicon-on-insulators) have made it possible for even more advancements. Nevertheless, as we approach both fundamental physical limits and practical restrictions, the energy efficiency advancements in nano-electronics has begun to decelerate recently[1–3]. The scalability of equivalent gate oxide thickness (EOT) and supply voltage (V_{DD}) of MOSFETs during the last 50 years highlight two historical patterns in Figure 4.1 which show how close we are to some of these constraints [4].

The effective oxide thickness is described as the thickness of silicon dioxide (SiO_2) required to obtain the equal capacitance as an oxide in the gate with a larger relative permittivity (ε_r) and thickness (d) since SiO_2 was initially employed as an oxide in the gate stack. As a result, EOT may be determined using the formula EOT = 3.9/ ($\varepsilon_r d$) having an SiO_2 relative permittivity of 3.9. While decreasing V_{DD} is the most efficient technique to enhance the integrated circuits efficiency, decreasing EOT is required in highly scaled devices to boost the electrostatic coupling of the gate to channel. Less voltage is required to turn the device on and off with a lower EOT because there is less of a drop in voltage across the gate oxide. For several decades, SiO_2 was utilized as an insulator material in MOSFETS based on silicon because of the superior quality of the Si/SiO_2 junction and its flexibility for fabrication. Quantum-mechanical electron tunneling through the gate dielectric increased to such a large extent beyond the

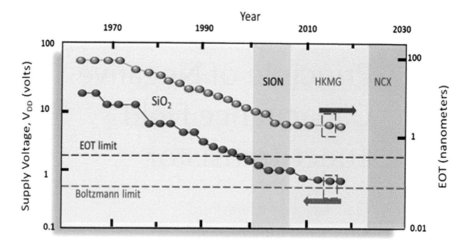

FIGURE 4.1   Historic patterns of voltage supply and EOT by year.

physical thickness of oxide of around 3 nm so that it began to considerably donate to the overall power dissipation [2,5–7]. As a result, at the start of the 20th century, nitrided SiO_2 gate dielectrics were used to elevate ε_r to about 5 and lower the EOT to below 2 nm. To maintain future scaling of EOT, nitrided SiO_2 (SiON) has to be superseded soon after the debut of the dielectrics with even higher ε_r [6].

The end result was the introduction of the so-called high-k metal gate (HKMG) technology in 2007 which relies on amorphous hafnium dioxide (HfO_2) dielectrics with ε_r approximately equivalent to 17, which enabled the scaling down of EOT to 0.8 nm in more recent devices [6]. However, EOT scaling down to 0.5 nm is essentially not conceivable without significantly reducing device performance and reliability due to the intrinsic presence of a thin SiO_2 interface layer between the high-k dielectric and the silicon channel [8]. Even though the EOT could be zero, a more basic constraint precludes further decreasing V_{DD}, which is required to increase the integrated circuit's efficiency. In order to simultaneously obtain adequate results [high ON current (I_{ON})] and static power usage [minimum OFF current (I_{OFF})] in any traditional FET, a high appropriate ratio of drain current (I_D) in the ON condition and in the OFF condition needs to be guaranteed. It has emerged that the fundamental lower bound for the gate voltage (V_G) required to modify the drain current at ambient temperature is 60 mV/decade owing to the Boltzmann distribution of carriers [9,10]. This 'Boltzmann limit' prevents the voltage supply from being decreased in traditional MOSFETs below about 0.5 V, which is not far from the 0.7 V employed in current technology, as shown in Figure 4.1.

Subthreshold swing is defined as the gate voltage needed to alter the current of a device by a decade in the subthreshold region. Mathematically, the subthreshold swing (SS) can be denoted as [9]:

$$SS = \frac{\partial V_G}{\partial \left(\log I_D\right)} = \frac{\partial V_G}{\partial \psi_s} \cdot \frac{\partial \psi_s}{\partial \left(\log I_d\right)} \tag{4.1}$$

$$SS = 2.3 * \frac{K_B T}{q} * \frac{\partial V_g}{\partial \psi_s} \qquad (4.2)$$

$$SS = 2.3 * \frac{K_B T}{q} * \left(1 + \frac{C_s}{C_{in}}\right) \qquad (4.3)$$

$$SS = 60\,\text{mV / decade} * \left(1 + \frac{C_s}{C_{in}}\right) \qquad (4.4)$$

where I_D is drain current, V_G is gate voltage, Ψ_s is the surface potential, T is the temperature, q is the charge, K_B is the Boltzmann constant, and C_{in} and C_s are the gate insulator and semiconductor capacitance, respectively. Several different approaches are currently being explored to address the Boltzmann tyranny issue of conventional transistors. Examples includes tunneling field effect transistors (TFETs) [11], nano-electromechanical switches (NEMs) [12], ionization metal oxide semiconductor transistors (IMOSs) [13,14], and negative field effect transistors (NCFETs) [10]. The first three methods adjust the transport mechanism, that is the manner in which electrons move within a transistor, in order to avoid hitting the 2.3 $K_B T/q$ minimum limit and normally these devices have a low ON current. On the other hand, in NCFET it has been theoretically demonstrated that it might be conceivable to preserve the transport mechanism while altering the electrostatic gating to raise the transistor's surface potential above what is normally feasible.

In Equation (4.4), C_s and C_{in} are normally positive numbers and therefore according to the capacitive voltage divider rule $\dfrac{\partial V_g}{\partial \psi_s}$ they will always be greater than 1. In 2008, Salahuddin and Datta originally suggested that if C_{in} is less than 0 in a traditional FET in Equation (4.3), one may circumvent the Boltzmann limit, which would allow further lowering of V_{DD} and, consequently, power dissipation of MOSFET [10]. In this circumstance, the negative capacitance (NC) would cause the surface potential to be internally amplified in relation to the applied gate voltage. Hence, in order to further enhance the energy efficiency of devices, one may circumvent both the looming EOT and the Boltzmann constraints in a negative capacitance field effect transistor (NCFET).

4.1.2 FERROELECTRICS

The discovery of switchable polarization behavior in Rochelle salt by Valasek in 1920 served as the catalyst for the development of ferroelectricity. To put it simply, a substance is considered to have ferroelectricity or to be a ferroelectric if it can manifest two possible equilibrium polarization states with opposing orientations even without an external electric field or spontaneous polarization [15]. Additionally, applying an electric field must be able to flip between these states. It is important to highlight that the ferroelectrics are a subset of a larger family of substances known as pyroelectrics that have a distinctive polar axis. Nevertheless, not all pyroelectrics may have the ability to reverse polarization, unlike ferroelectrics. The spontaneous polarization in crystalline ferroelectrics results from an asymmetry in the ionic configurations within a unit cell.

Ferroelectrics are a unique family of materials that exhibit polarization even in the absence of an electric field. It is possible to reverse this spontaneous polarization, which results from a noncentrosymmetric structure, by applying an electric field that is stronger than the coercive field [15]. Ferroelectrics are perfect for nonvolatile memory systems since this typically results in polarization hysteresis [16]. Even though each ferroelectrics are both piezoelectric and pyroelectric, these substances are utilized in devices like detectors (infrared) and ultrasonic transducers. Further, ferroelectrics are advantageous in microwave electronics because their permittivity is dependent on the applied electric field. In addition, the ferroelectric permittivity can even turn negative under specific circumstances. The ferroelectric changes itself to paraelectric with zero spontaneous polarization above the Curie temperature.

Most ferroelectrics have a crystalline structure of perovskite material. Take $BaTiO_3$ as an example of a ferroelectric of the perovskite (ABO3) type [17]. The usual stoichiometry of the perovskite structure is ABO3, where cations are 'A' and 'B' and 'O' is an anion. At high temperatures, it displays a cubic form, as seen in Figure 4.2(a). In this instance, the unit cell carries a satisfying centrosymmetry. As a result, in this condition, the net polarization is zero, and is also known as the paraelectric phase. As seen in Figures 4.2(b) and 4.2(c), it goes through a phase transition and becomes a tetragonal unit cell when the temperature is lowered below 120°C. Curie temperature is a term used to describe the transition temperature. A permanent dipole moment is created when the central Ti atom in the new tetragonal phase is adjusted either upwards or downwards along the c-axis. In general, this attribute can be acquired through any one of the six orientations along the coordinate axes. The entire ferroelectric volume breaks into domains, which are areas of uniform polarization, in order

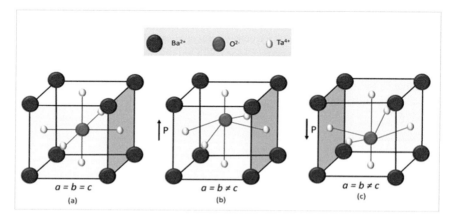

FIGURE 4.2 $BaTiO_3$ structure: (a) for temperatures greater than Curie temperature (cubic phase); (b) and (c) for temperatures less than Curie temperature (tetragonal phase) with two different polarization orientations.

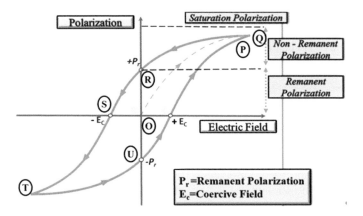

FIGURE 4.3   Polarization–electric field hysteresis curve.

to decrease the ferroelectric material energy for specific electrical and mechanical boundary conditions. Domain walls are the physical barriers that divide the domains.

According to Figure 4.3, the polarization exhibits a hysteresis characteristic when an electric field is applied to a ferroelectric material [18]. It also fluctuates nonlinearly under this condition. Consequently, the net polarization is zero, as indicated by point O in Figure 4.3, in the primarily formed ferroelectric sample, since all of the domains are initially randomly orientated. Domains begin to align and polarization intensifies along the path OP as the electric field increases. The polarization begins to saturate as the electric field increases from P to Q. All the domains align in the same direction at Q. In simple terms, the ferroelectric in this condition exhibits single domain behavior. Due to the fact that not all domains are misaligned, the removal of the electric field from Q to R does not reduce the polarization to zero. The term 'remnant polarization' refers to the polarization value at point R. A coercive field (E_c), also known as an opposing electric field of quantity, must be applied in order to reduce the polarization to zero. Domains are aligned in the opposite manner from S to T as the electric field increases further. The ferroelectric follows the path T to U to Q, causing a hysteresis loop if the electric field is swept forward once again [18]. As a result, the ferroelectric develops a memory since the polarization value at any given electric field depends on the material's history. In general, the physical technique of a ferroelectric's polarization switching dynamics is a complex process that involves the formation of new domains and their continuous expansion in response to an opposite electric field.

One of the most researched classes of ferroelectrics is perovskite. Additional examples of this group are bismuth ferrite ($BiFeO_3$), lead titanate ($PbTiO_3$), lead zirconium titanate ($PbZr_xTi_{1-x}O_3$), barium strontium titanate ($BaSr_xTi_{1-x}O_3$), and recently discovered doped HfO_2-based ferroelectrics such as hafnium zirconium oxide ($Hf_xZr_{1-x}O_2$), silicon-doped hafnium oxide, and hafnium aluminum oxide [19–21]. Due to their compatibility with the CMOS manufacturing process, HfO_2-based ferroelectrics are extremely important for the semiconductor industry. As a result, it is now possible to incorporate a ferroelectric layer inside a traditional FET for applications like memory and extremely low-power switching [22,23]. In addition to these inorganic ferroelectrics, polyvinylidenefluoride-trifluoroethylene $P(VDF_x\text{-}TrFE_{1-x})$ copolymer is a significant organic ferroelectric material. $Trans(CH_2CF_2)_x(CH_2\text{-}CHF)_{1-x}$ molecular chains create molecular dipoles in this material, which causes ferroelectricity [24].

4.1.3 NEGATIVE CAPACITANCE

In 2008, Salahuddin and Datta described a plan to navigate the impassable negative slope section [10]. They recommended adding a second dielectric layer in series with the ferroelectric so that the system's overall energy has a minimum at $P = 0$, as shown in Figure 4.4. In order to keep the total curvature positive, the dielectric must have a positive curvature that is sufficiently large to counteract the ferroelectric's negative curvature. A device that stores charge is known as a capacitor. The rate at which a charge Q increases with voltage V is known as a device's capacitance ($C = \dfrac{dQ}{dV}$). Therefore, by concept, for a negative capacitor, Q drops as V increases. Capacitance can also be understood in the context of the free energy U. The energy landscape of a negative capacitor is a reversed parabola. $U = \dfrac{Q^2}{2C}$ applies to a linear capacitor. The capacitance can be explained in terms of free energy using the following definition:

$$C = \frac{1}{\left(\dfrac{d^2U}{dQ^2}\right)} \tag{4.5}$$

The nonlinear capacitor is also subject to the same relationship. In other words, a negative capacitance corresponds to the region of negative curve in an insulating material's energy landscape.

In order for negative capacitance to be stable, the following conditions must be fulfilled:

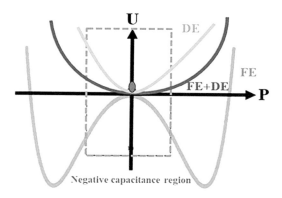

FIGURE 4.4   Energy landscapes of a dielectric, a ferroelectric, and their series arrangement. The NC section of the ferroelectric energy landscape is displayed inside the rectangular box.

$$\frac{1}{Total\,capacitance\,(C_T)} = \frac{1}{Dielectric\,capacitance\left(C_D\right)}$$
$$+ \frac{1}{-\left|Ferroelectric\,capacitance\left(C_F\right)\right|} > 0 \qquad (4.6)$$

$$\Rightarrow \left|C_F\right| > C_D \qquad (4.7)$$

Where C_D, C_F, and C_T represent the dielectric, ferroelectric, and total capacitances, respectively. A hysteric behavior is the outcome of violating the criterion in Equation (4.6).

4.2 BASIC OPERATION PRINCIPLES OF NEGATIVE CAPACITANCE FIELD EFFECT TRANSISTORS

The negative capacitance effect makes it possible to overcome the Boltzmann limitation and supported steep switching operation in ferroelectric materials. As can be seen in Figure 4.5(a), the ferroelectric material is incorporated into the gate stack of a regular MOSFET to construct a negative capacitance FET (NCFET) [18]. The basic MOS arrangement behaves as the dielectric that can maintain the ferroelectric's negative capacitance condition. The ferroelectric therefore produces an amplification in internal voltage because of the negative capacitance characteristic of ferroelectric, which can be determined from the equivalent voltage divider circuit of the NCFET presented in Figure 4.5(b). Ferroelectrics are advantageous for tunable capacitors in microwave electronics because their permittivity is dependent on the applied electric field. The ferroelectric permittivity may even turn negative in some circumstances. The ferroelectric changes into a paraelectric phase above the Curie temperature, where spontaneous polarization is zero.

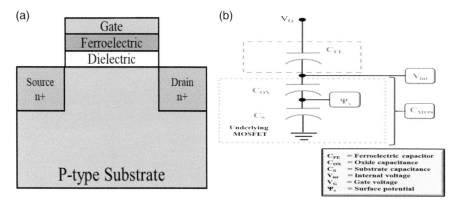

FIGURE 4.5   (a) Structure of NCFET; (b) circuit model of NCFET.

In linear dielectrics, electric polarization is directly related to the electric field, leading to the displacement field $(D) = \varepsilon_o \varepsilon_r E$. The ferroelectric permittivity can be realized from the displacement field formula as:

$$D = \epsilon_o \cdot E + P \tag{4.8}$$

$$\varepsilon_o \cdot \varepsilon_f = \frac{dD}{dE_F} = \varepsilon_o + \frac{dP_s}{dE_F} \approx \frac{dP_s}{dE_F} \tag{4.9}$$

where E_F represents the ferroelectric's electric field and P_s is spontaneous polarization. Thus, a ferroelectric's nonlinear permittivity and, consequently, capacitance would be negative when the polarization in the ferroelectric changes in the opposite direction of the electric field. The possibility of a negative capacitance as a result of ferroelectric polarization instability was initially foreseen by Landauer in 1976. The justification was established on a straightforward thermodynamic model developed by Landauer known as the Landau model [25], in which the free energy of ferroelectric can be represented as

$$U_F = \alpha \cdot P_S^2 + \beta \cdot P_S^4 - E_F P_S \tag{4.10}$$

where α and β are Landau coefficients and below Curie temperature alpha is less than zero and beta is greater than zero. Figure 4.6(a) presents the free energy versus polarization landscape at $E_F = 0$, considering a second-order phase transition. The maxima at $P_s = 0$ represents an unstable state, whereas the two minimum energy levels indicate a stable polarization condition [26]. The equation for the electric field is obtained by minimizing U_F with respect to P_s in Figure 4.6(b) as:

$$E_F = 2 \cdot \alpha \cdot P_s + 4 \cdot \beta \cdot P_S^3 \tag{4.11}$$

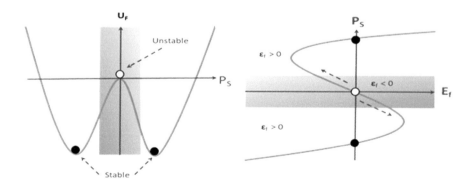

FIGURE 4.6   (a) Polarization versus free energy; (b) polarization versus electric field.

Since α <0, it is clear that the permittivity should always be negative for low values of P_s, resulting in an 'S'-shaped negative slope and a curve (negative) for the energy landscape [27]. It is instantly obvious from Figure 4.6(a) that the area having negative permittivity (i.e., negative capacitance) is unstable thermodynamically because if the ferroelectric is taken in isolation, it relates directly to a maximal free energy. There are two approaches to access the negative capacitance region: (1) If we apply an electric field to flip the polarization of the ferroelectric from one stable condition to another, transient negative capacitance should be observed during switching for a brief time period [28]. (2) We include the ferroelectric into a bigger structure so that, when the system's total free energy is minimized, the negative capacitance condition becomes stable thermodynamically.

4.2.1 METAL FERROELECTRIC METAL (MFM) CAPACITOR

Consider an ideal metal ferroelectric metal (MFM) capacitor which is transitioning through a negative to a positive polarization condition shown in Figure 4.7(a). For varying stages in the switching operation, the energy landscapes of the ferroelectric are depicted in Figures 4.7(b)–4.7(f). When the external voltage is zero, the landscape is firstly symmetric and polarization is negative. As shown in Figure 4.7(c), a voltage

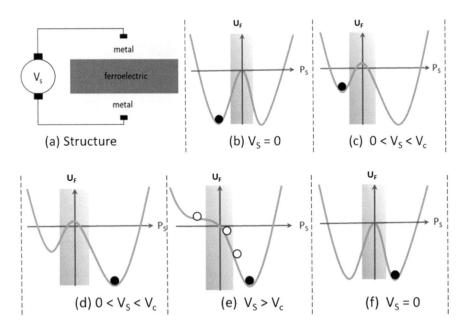

FIGURE 4.7 Switching process of a MFM capacitor: (a) structure of MFM at given voltage V_S; (b) free landscape energy at $V_S = 0$ (assuming spontaneous polarization [P_S] to be negative); (c) providing $V_S < V_C$; (d) spontaneous polarization switching from negative to positive at $V_S > V_C$; (e) P_S remains positive when $V_S <$ zero (f).

greater than zero will cause the curve to tilt to the right side, lifting the energy minima of the left side and decreasing the barrier. When the given voltage is greater than the critical voltage (V_C), the barrier disappears and polarization passes over the negative capacitance area till it hits the appropriate minima, as shown in Figure 4.7(d). Further, as the external voltage is reduced below the coercive voltage, the energy barrier will reappear, keeping the polarization in the right minima, as illustrated in Figures 4.7(e) and 4.7(f). The MFM capacitor, however, will constantly remain in a positive capacitance condition when switching takes place in the ferroelectric because of the unstable nature of the NC region [27].

In these circumstances, it is obvious that the polarization will usually rise with time [(dP_S/dt) greater than 0]. As a result, during switching in the transient negative capacitance (TNC) condition, the voltage and, consequently, ferroelectric field must drop with time [(dE_F/dt) less than 0]. As P_S is rising with time, it is clear that a transient negative effect will result if P_S changes more quickly than the metal free charge. One approach to witness this TNC experimentally is to apply a voltage through the serial construction between a metal ferroelectric metal capacitor and an additional circuit component, such as a resistor, a dielectric capacitor, or even the gate of a MOSFET [29]. The added circuit component will slow down the flow of screening charge to the MFM capacitor from an external circuit while simultaneously allowing the voltage of ferroelectric to decline with time while the given voltage rises [30]. In addition, there are two significant drawbacks to using transient negative capacitance in applications: (1) Transient negative capacitance is generally linked to an enormous hysteresis, which is harmful for energy-efficient applications because it causes dissipation of power and demands an increment in the supplied voltage [31]. (2) According to the established trade-off between voltage and time, transitioning a ferroelectric entirely through one stable polarization phase to the other can only be accomplished rapidly with a high voltage or slowly require with a low voltage, not both at the same time[32,33]. Thus, transient negative capacitance effects are therefore too sluggish for higher speed circuits that operate at lower voltage and giga hertz frequencies [34]. These two limitations are a primary outcome of the instability of the negative capacitance state. Theoretically, we could, however, get beyond both of these restrictions if we could stabilize the inherent NC state [35].

4.2.2 METAL FERROELECTRIC INSULATOR METAL CAPACITOR

Salahuddin and Datta first proposed that a positive capacitance may be introduced in series with the ferroelectric capacitance to stabilize the ferroelectric NC condition at $P_S = 0$[10]. As a result, there would be zero polarization hysteresis and the system's overall capacitance will be positive. It has recently been established that the impact of domain creation and leakage current, which destabilize the negative capacitance state, make a stabilization like this implausible when an MFM capacitor is used serially with a metal dielectric metal capacitor [36,37]. Thus, there shouldn't be a metal layer between the positive capacitance layer and the ferroelectric layer [36,38,39].

To comprehend the fundamentals of NC stabilization, the free energy concept is utilized for metal ferroelectric insulator metal capacitor configuration. The dielectric's free energy with the electric field (E_D) and relative permittivity (ε_D) can be defined as:

$$U_D = \left(\frac{D^2}{2 \cdot \varepsilon_0 \cdot \varepsilon_D} \right) - \left(E_D \cdot D \right) \qquad (4.12)$$

Assuming D is approximately equal to P_S, then the total free energy (/area) can be expressed as follows:

$$U_T = U_F T_F + U_D T_D \qquad (4.13)$$

$$U_T = \left(\alpha \cdot T_F + \frac{T_D}{2 \cdot \varepsilon_0 \cdot \varepsilon_D} \right) \cdot P_S^2 + T_F \cdot \beta \cdot P_S^4 - V \cdot P_S \qquad (4.14)$$

where T_D and T_F are dielectric thickness and ferroelectric thickness, respectively. The ferroelectric negative capacitance phase would be stabilized when the part in parentheses in Equation (4.12) turns positive, because the curve of U_T with reference to P_S is always positive and, therefore, the overall capacitance will be positive, which can be defined as follows:

$$T_F \leq T_{F,critical} = -\frac{T_D}{2 \cdot \alpha \cdot \varepsilon_0 \cdot \varepsilon_D} \qquad (4.15)$$

Nevertheless, particularly without the $D \leftrightarrow P_S$ assumption, $T_{F, critical}$ is indirectly proportional to the dielectric capacitance [36]. Now consider the situation when T_F is greater than $T_{F, critical}$ (for instance, if the insulator layer is much thinner), where free energy for various voltages are presented in Figures 4.8(a)–4.8(f), presuming that P_S is primarily negative. The ferroelectric is depolarized by the dielectric in comparison to the ideal metal ferroelectric metal scenario in Figure 4.7, which lowers the energy barrier in U_T and reduces spontaneous polarization. The ferroelectric maintains a positive capacitance at $V_S = 0$ in Figure 4.8(b). Unscreened polarization causes a positive field of depolarization in the ferroelectric and an opposing negative field in the insulator. The depolarization phenomenon causes the ferroelectric field to significantly drop and P_S shifts to the positive direction when V_S is raised above V_C as shown in Figure 4.8(d), producing a transient negative capacitance as in Figure 4.7. After that, when V_S is zeroed out, P_S remains in the positive energy minima, as shown in Figures 4.8(e) and 4.8(f).

Figures 4.9(a)–4.9(c) depict the interesting situation where T_F is less than or equal to $T_{F, critical}$. In this situation, the stack's total free energy is reduced at $V_S = 0$, which causes the negative capacitance state at $P_S = 0$ to become stable [Figure 4.9(b)]. The maxima of U_F and the minima of U_D and U_T now coincide. When a positive voltage is given [Figure 4.9(c)], P_S rises while E_F falls as a result of the dielectric's depolarization field, so that $dP_S/dE_F < 0$. Although the curve of U_T (hence the total capacitance) will always be positive, this overall system is stable [39]. The ferroelectric will transition into a condition of positive capacitance when V_S is raised over a critical voltage V_C [Figure 4.9(d)]. However, after the voltage is subsequently decreased to zero, the

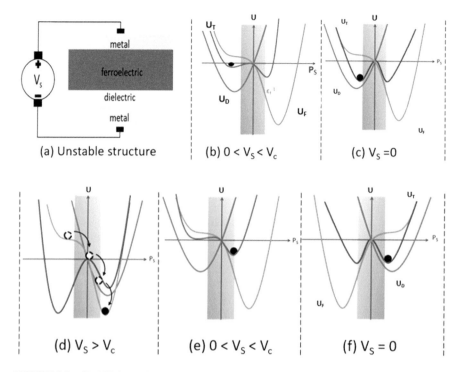

FIGURE 4.8 Stabilizing NC in a ferroelectric-dielectric hetero-structure with V_S. (a) Negative capacitance is not stable if the insulator layer is very thin. Switching from negative P_S from (b) $V_S = 0$ and increasing V_S (c) more than V_C (d). Corresponding reduction of V_S (e) again results in a positive P_S at $V_S = 0$ (f) for a thick insulator.

ferroelectric reverts to its earlier negative capacitance condition [Figures 4.9(e) and 4.9(f)]. In the entire range of operation, there are negligible instabilities (no negative curve of U_T), hence no hysteresis is anticipated. This runs contrary to the previously stated temporary NC effect.

There are additional effects to take into account when switching from the simpler capacitors to NCFET devices, as illustrated in Figures 4.8 and 4.9. First, as the semiconductor channel transitions from the depletion to the inversion zone, the stabilizing positive capacitance is very nonlinear. As a result, the thickness of the ferroelectric range where negative capacitance can be stabilized through the entire range of operation is substantially lowered. The utilization of reduced density of states channel materials, as mentioned by Cao and Banerjee, may be advantageous for the configuration of NCFET devices [40]. Secondly, particularly in short channel devices with inhomogeneous channel potential, the drain voltage can have a local effect on the NC stability. Therefore, domains may arise and the ferroelectric polarization will also become irregular. As a result, domains may arise and the ferroelectric polarization will also become irregular. However, simulations show that even at large drain voltages, hysteresis-free stable NC is still feasible [41]. It has been demonstrated that

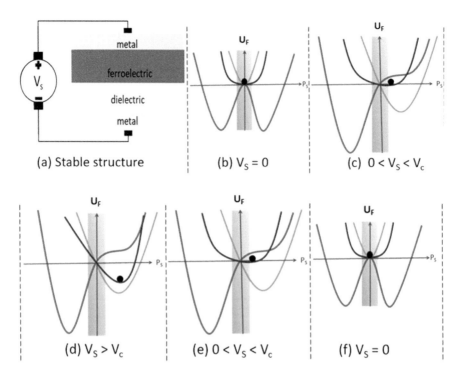

FIGURE 4.9 Stabilizing NC in a ferroelectric-dielectric hetero-structure with V_S. (a) Negative capacitance phase is stabilized by the energy of dielectric at $V_S = 0$. (b) For small given voltage, negative capacitance is still stable (c) above V_C, the capacitance of ferroelectric becomes positive (d) reducing V_S (e) the ferroelectric comes to the stable negative capacitance phase (f).

rising drain voltage in the NCFET causes the channel potential to decrease, which results in negative drain-induced barrier lowering and negative differential resistance in the output characteristics [42,43].

The ferroelectric negative capacitance has been demonstrated to be an unstable condition in preceding sections, and in a secluded ferroelectric capacitor, the NC phases can only be addressed in a time-dynamic manner. This section outlines how the addition of a positive capacitor serially stabilizes the ferroelectric NC condition. The studies in this part are significant because stabilizing the ferroelectric negative capacitance is necessary to decrease the subthreshold swing in a field effect transistor. The effective capacitance is increased relative to the respective dielectric as an outcome of the stabilization of the ferroelectric NC.

Figure 4.10 illustrates a schematic of the P–E curve and channel charge of NCFET. To determine the operating point of a ferroelectric MOS capacitor, the polarization and channel charge density must coincide. Therefore, the point where the P–E curve and the channel charge load line cross is used to determine the static operating point of NCFET as depicted in Figure 4.10. Ferroelectric MOS transistors function as NCFETs when the P–E curve has a mono crossing point in the negative capacitance

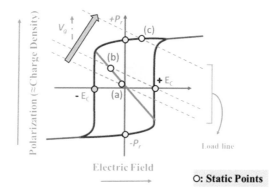

FIGURE 4.10 Polarization–electric field (P–E) and load line of channel.

region. In contrast to a traditional MOSFET, when gate voltage (V_g) is provided the NCFET produces a negative oxide field on the insulator of gate. The operating point (a) in Figure 4.10 shifts to (b) when the gate voltage is increased and the oxide field (negative) enlarges even more. As illustrated in Figure 4.10, the band tilts more than the supplied gate voltage. As a result, the NCFET is able to generate a larger current than traditional MOSFETs because the potential of the surface is efficiently amplified against the gate voltage and more charges are produced in the channel. The operating point now shifts to (c) as the gate voltage is increased more and the field of oxide switches, and finally the band diagram resembles that of typical MOSFETs. The charge density of the channel in the MOS capacitor is lesser in the subthreshold area and rises by $\sqrt{\Psi_s}$, whereas it is large in the strong inversion area and rises by $e^{(q\Psi_s/2kT)}$. The density of charge in the channel needs to be large enough to meet the ferroelectric polarization in the strong inversion area instead of the subthreshold area in order to utilize the negative oxide voltage. Due to this, the threshold voltage decreases and a steeper slope is mostly seen beyond the threshold region.

4.3 CONCLUSION

In this chapter, the basic NCFET structure, its characteristics, and operation with its stabilization process are described. To overcome the Boltzmann constraint and allow steep switching action, the ferroelectric negative capacitance phenomenon can be utilized. A NCFET has basically the same structure as a conventional MOSFET, only a ferroelectric material is incorporated in the gate stack. The basic MOS configuration serves as the dielectric that can help in stabilization of the ferroelectric's negative capacitance condition. The ferroelectric therefore provides an amplification in the internal voltage as a result of its negative capacitance characteristic that can be determined from the corresponding voltage-divider circuit of an NCFET. Besides lowering SS, the voltage gain can be utilized to enhance the charge density of the channel in the inversion operation region, which results in a higher ON current in NCFETs compared to traditional MOSFETs. Additionally, various advantages of using NCFET devices include having lower subthreshold slope, negligible hysteresis, improved ON current, and reduced subthreshold swing. In view of the aforementioned parameters, the NCFET is a promising replacement for the MOSFET for high-speed, cost-effective, and ultra-low-power circuits. Numerous research works have backed the use of NCFET as an ultra-low-power transistor in the future. However, for commercial production, there are various challenges that must be resolved such as (a) the solution for a HfO_2-based, 1-nm ferroelectric layers in particular, with appropriate controllability of imperfections and thickness, (b) the in-depth explanation of random variations of NCFET, and (c) the process instructions for capacitance matching in wafer-level procedures. Finally, NCFET has garnered outstanding results in numerous applications such as memories, ultra-low power devices, inverters, the Internet of Things, radio frequency sensing, analog circuits, filters, and biomedical devices.

REFERENCES

[1] A. Danowitz, K. Kelley, J. Mao, J. P. Stevenson, and M. Horowitz (2012) CPU DB. *Commununications of the ACM* 55(4): 55–63.
[2] J. Koomey, S. Berard, M. Sanchez, and H. Wong (2011) Implications of historical trends in the electrical efficiency of computing. *IEEE Annals of the History of Computing* 33(3): 46–54.
[3] R. D. Isaac (2022) Reaching the limits of CMOS technology. In: *IEEE 7th Topical Meeting on Electrical Performance of Electronic Packaging (Cat. No.98TH8370)*, no. 9 14, p. 3.
[4] S. E. Thompson, R. S. Chau, T. Ghani, K. Mistry, S. Tyagi, and M. T. Bohr (2005) In search of 'forever,' continued transistor scaling one new material at a time. *IEEE Transactions on Semiconductor Manufacturing* 18(1): 26–36.
[5] G. E. Moore (2003) No exponential is forever: but 'Forever' can be delayed! [semiconductor industry]. *2003 IEEE International Solid-State Circuits Conference, 2003. Digest of Technical Papers. ISSCC* 1: 20–23.
[6] B. H. Lee, J. Oh, H. H. Tseng, R. Jammy, and H. Huff (2006) Gate stack technology for nanoscale devices. *Materials Today* 9(6): 32–40.
[7] C. Hu, D. Chou, P. Patel, and A. Bowonder (2008) Green transistor – a V_{DD} scaling path for future low power ICs. *2008 International Symposium on VLSI Technology, Systems and Applications (VLSI-TSA)* 84(10): 14–15.
[8] L.-Å. Ragnarsson, T. Chiarella, M. Togo, T. Schram, P. Absil, and T. Hoffmann (2011) Ultrathin EOT high-κ/metal gate devices for future technologies: Challenges, achievements and perspectives (invited). *Microelectronic Engineering* 88(7): 1317–1322.
[9] D. C. Northrop (1986) Book review: Semiconductor devices—Physics and technology. *International Journal of Electrical Engineering Education* 23(1): 64–64.
[10] S. Salahuddin and S. Datta 2008) Use of negative capacitance to provide voltage amplification for low power nanoscale devices. *Nano Letters* 8(2): 405–410.
[11] A. M. Ionescu and H. Riel (2011) Tunnel field-effect transistors as energy-efficient electronic switches. *Nature* 479(7373): 329–337.
[12] Hei Kam, D. T. Lee, R. T. Howe and Tsu-Jae King (2005) A new nano-electro-mechanical field effect transistor (NEMFET) design for low-power electronics. *IEEE International Electron Devices Meeting, 2005. IEDM Technical Digest* 2005(c): 463–466.

[13] S. Ramaswamy and M. J. Kumar (2014) Junctionless impact ionization MOS: Proposal and investigation. *IEEE Transactions on Electron Devices* 61(12): 4295–4298.

[14] Woo Young Choi, Jae Young Song, Jong Duk Lee, Young June Park and Byung-Gook Park (2005) 100-nm n-/p-channel I-MOS using a novel self-aligned structure. *IEEE Electron Device Letters* 26(4): 261–263.

[15] L. E. Cross and R. E. Newnham (1987) History of ferroelectrics. *Ceramics and Civilization* III: 289–305.

[16] T. Mikolajick, U. Schroeder and S. Slesazeck (2020) The Past, the present, and the future of ferroelectric memories. *IEEE Transactions on Electron Devices* 67(4): 1434–1443.

[17] R. Landauer (1957) Electrostatic considerations in $BaTiO_3$ domain formation during polarization reversal. *Journal of Applied Physics* 28(2): 227–234.

[18] B. Choudhuri Malvika and K. Mummaneni (2022) A review on a negative capacitance field-effect transistor for low-power applications. *Journal of Electronic Materials* 51(3): 923–937.

[19] O. Auciello (1997) A critical comparative review of PZT and SBT – based science and technology for non-volatile ferroelectric memories. *Integrated Ferroelectrices* 15(1–4): 211–220.

[20] R. R. Das, S. B. Majumder and R. S. Katiyar (2002) Comparison of the electrical characteristics of PZT and SBT thin films. *Integrated Ferroelectrics* 42(1): 323–334.

[21] T. S. Böscke et al. (2011) Phase transitions in ferroelectric silicon doped hafnium oxide. *Applied Physics Letters* 99(11): 112904.

[22] S. Mueller et al. (2013) From MFM capacitors toward ferroelectric transistors: Endurance and disturb characteristics of HfO_2-based FeFET devices. *IEEE Transactions on Electronic Devices* 60(12): 4199–4205.

[23] T. S. Böscke, J. Müller, D. Bräuhaus, U. Schröder and U. Böttger (2011) Ferroelectricity in hafnium oxide thin films. *Applied Physics Letters* 99(10): 102903.

[24] R. G. Kepler and R. A. Anderson (1992) Ferroelectric polymers. *Advances in Physics* 41(1): 1–57.

[25] O. F. Sound, N. A. Second, and P. T. Point (1965) On the anomalous absorption of sound near a second order phase transition point. *Collected Papers of L. D. Landau* 469(4): 626–629.

[26] I. Luk'yanchuk, Y. Tikhonov, A. Sené, A. Razumnaya, and V. M. Vinokur (2019) Harnessing ferroelectric domains for negative capacitance. *Communications Physics* 2(1): 22.

[27] M. Hoffmann, S. Slesazeck and T. Mikolajick (2021) Progress and future prospects of negative capacitance electronics: A materials perspective. *APL Materials* 9(2): 020902.

[28] A. I. Khan et al. Negative capacitance in a ferroelectric capacitor. *Natural Materials* 14(2): 182–186.

[29] S.-C. Chang, U. E. Avci, D. E. Nikonov, S. Manipatruni and I. A. Young (2018) Physical origin of transient negative capacitance in a ferroelectric capacitor. *Physical Review Applied* 9(1): 014010.

[30] Y. J. Kim et al. (2017) Voltage drop in a ferroelectric single layer capacitor by retarded domain nucleation. *Nano Letters* 17(12): 7796–7802.

[31] M. Hoffmann et al. (2018) Ferroelectric negative capacitance domain dynamics. *Journal of Applied Physics* 123(18): 184101.

[32] A. I. Khan et al. (2016) Negative capacitance in short-channel FinFETs externally connected to an epitaxial ferroelectric capacitor. *IEEE Electron Device Letters* 37(1): 111–114.

[33] L. Pintilie et al. (2020) Polarization switching and negative capacitance in epitaxial Pb Zr$_{0.2}$ Ti$_{0.8}$ thin films. *Physical Review Applied* 14(1): 014080.

[34] Z. C. Yuan et al. (2016) Switching-speed limitations of ferroelectric negative-capacitance FETs. *IEEE Transactions on Electron Devices* 63(10): 4046–4052.

[35] K. Chatterjee, A. J. Rosner and S. Salahuddin (2017) Intrinsic speed limit of negative capacitance transistors. *IEEE Electron Device Letters* 38(9): 1328–1330.

[36] M. Hoffmann, M. Pešić, S. Slesazeck, U. Schroeder and T. Mikolajick (2018) On the stabilization of ferroelectric negative capacitance in nanoscale devices. *Nanoscale* 10(23): 10891–10899.

[37] Malvika, B. Choudhuri and K. Mummaneni (2022) A new pocket-doped NCFET for low power applications: Impact of ferroelectric and oxide thickness on its performance. *Micro and Nanostructures* 169(August): 207360.

[38] A. I. Khan, U. Radhakrishna, K. Chatterjee, S. Salahuddin and D. A. Antoniadis (2016) Negative capacitance behavior in a leaky ferroelectric. *IEEE Transactions on Electron Devices* 63(11): 4416–4422.

[39] T. Rollo, F. Blanchini, G. Giordano, R. Specogna and D. Esseni (2020) Stabilization of negative capacitance in ferroelectric capacitors with and without a metal interlayer. *Nanoscale* 12(10): 6121–6129.

[40] W. Cao and K. Banerjee (2020) Is negative capacitance FET a steep-slope logic switch? *Nature Communications* 11(1): 196.

[41] Z. Dong and J. Guo (2017) A simple model of negative capacitance FET with electrostatic short channel effects. *IEEE Transactions on Electron Devices* 64(7): 2927–2934.

[42] H. Agarwal et al. (2018) Engineering negative differential resistance in NCFETs for analog applications. *IEEE Transactions on Electron Devices* 65(5): 2033–2039.

[43] J. Seo, J. Lee and M. Shin (2017) Analysis of drain-induced barrier rising in short-channel negative-capacitance FETs and its applications. *IEEE Transactions on Electron Devices* 64(4): 1793–1798.

5 Basic Operational Principle of Anti-ferroelectric and Ferroelectric Materials

Umesh Chandra Bind[a],
Shiromani Balmukund Rahi[b]
[a]Institute of Engineering and Technology,
Dr.A.P.J. Abdul Kalam Technical University, Lucknow 226031
[b]India Department of Electrical Engineering, Indian Institute
of Technology Kanpur, 208016, India

5.1 INTRODUCTION

So far, there has been a continuous and steady growth of the market for mobile computing devices, wireless systems, and biomedical implantable devices [1–5]. This has led to a consistent growth in the demand for low-power circuits and systems. The current CMOS technology could not fulfill these requirements owing to the non-scalability of the threshold voltage (V_{TH}) due to immutable subthreshold swing (SS), leading to greater leakage of current for low-power circuit operation [6–8]. With the recent advent of semiconductor technology, the performance of conventional FETs has been steadily improved by scaling down the feature size. However, the operating voltage cannot be scaled down with the channel length scaling because the subthreshold slope (SS) cannot be reduced below 60 mV/decade (Boltzmann tyranny) [9–15]. The demand for low-power VLSI circuits and systems has been fulfilled by MOSFET by continuous scaling of supply voltage (V_{DD}) and technology scaling [16–20]. The scaling of MOSFET have been the critical driving factor in improving the energy efficiency. However, supply voltage (V_{DD}) scaling is limited to the near threshold (V_{TH}) region of classical MOSFET [21–24]. Reducing V_{DD} much below V_{TH} leads to a significant loss in speed and energy efficiency [25–29]. This phenomenon is directly related to the fundamental limit of 60 mV/decade on the SS at room temperature (300 K) for MOSFETs. Surpassing and overcoming this limit could enable more aggressive V_{DD} scaling, leading to ultra-low-power dissipation. Therefore, devices that exhibit steep switching characteristics (i.e., SS < 60 mV/decade at room temperature) are being actively and widely explored [13–19]. Several steep slope devices have emerged in the recent past owing to their promise

DOI: 10.1201/9781003373391-5

to deliver higher ON-state current at lower voltages compared to the standard MOSFETs [30–37].

5.2 NEGATIVE CAPACITANCE

In general, capacitance is the ratio of charge to voltage in any material. In other words, capacitance is the charge store/hold per unit volt in any material. This feature can be identified when material is kept between two electrodes that are suitable to tune the applied voltage across the material (Figure 5.1). The capacitance of the material is given by:

$$C = \frac{Q}{V} \qquad (5.1)$$

where, C, Q, and V represent capacitance, charge, and applied voltage, respectively. It is one phase or very usual feature of materials which reflect that as one increases the applied voltage across the material it accumulates more and more charge inside the material and vice versa. Now, we can talk about the effects of very small changes in applied voltage resulting in the charge accumulation feature of the material which is preferably applicable in the operation of low-power devices and can be express as:

$$C = \frac{dQ}{dV} \qquad (5.2)$$

where, dQ is small changes in the charges on a very small applied voltage dV. In materials, this capacitance reflects different features at different scales, and especially at nanoscale dimensions it become more prominent owing to its size, shape, stoichiometric composition, phases, etc.

The capacitance feature/behavior can be also analyzed by calculating the total energy stored in the material sandwich between metals/electrodes and can be expressed as:

$$\text{stored energy (U)} = \frac{Q^2}{2C}$$

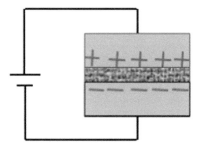

FIGURE 5.1 Schematic diagram of a parallel plate capacitor with dielectric material.

Therefore, the capacitance can be expressed in terms of stored energy as:

$$C = \frac{1}{\left(\dfrac{d^2U}{dQ^2}\right)} \tag{5.3}$$

This is the inverse of second-order differentiation of stored energy with respect to charges. In general, it very well known that as voltages increase the accumulated charge increases, as shown in Figure 5.2(a) and its respective variation of the stored energy with respect to an increase in the charges reflects the molecular phase stability inside the material [Figure 5.2(b)]. This type of material behavior varies between materials.

Some materials have unusual phases where the accumulation of charges increases with a decrease in the applied voltage across the materials and vice versa [Figure 5.2(c)]. In such cases, the stored energy is also a function of charge accumulation but it reflects an inverted parabolic shape that means the molecular phase of this type of materials is unstable [Figure 5.2(d)]. In other words, we can say that it

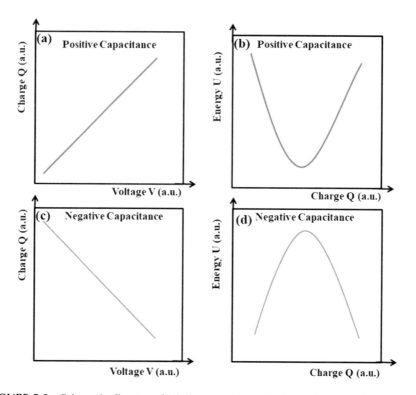

FIGURE 5.2 Schematic diagram of (a) charge-voltage of a dielectric material, (b) energy-charge response of a dielectric material, (c) charge-voltage of a negative capacitive material, and (d) energy-charge response of a negative capacitive material.

reflects a quick transitional situation in the phase of the material. This charge storage phenomenon/feature is known as the negative capacitance of the material.

To understand this, let's consider a simple example of a parallel plate capacitor as shown in Figure 5.1, where, when the voltage is applied across the material's two opposite surfaces, the capacitance can be calculated by

$$C = \frac{\varepsilon_0 \varepsilon_r A}{t} \tag{5.4}$$

where, ε_0, ε_r, A, and t are vacuum permittivity, relative permittivity, area of one side surface, and thickness of the material, respectively. As voltage is applied, an electric field is developed across the surface leading to the creation of an additional field inside the material known as the displacement field, which is given by

$$D = \varepsilon_0 \varepsilon_r E + P \tag{5.5}$$

where, D, E, and P represent the displacement field, electric field development through the applied voltage, and electric polarization, respectively.

The displacement field is dependent on the relative permittivity and polarization feature of the material. Polarization is the induced dipole moment per unit volume, and greatly depends on the features of the atoms or molecules available inside the material. Therefore, it varies from material to material as well as with material size, and such features are generally reflected in the dielectric material owing to its tightly bound positive and negative charges. On application of the electric field when it is polarized is shown in Figure 5.3, in which polarization is directly proportional to the electric field, and given by

$$P = N\alpha E \tag{5.6}$$

where, N and α are the number of atoms per unit volume and atomic/molecular polarizability proportionality constant. Thus, a small change in the electric field leads to variation in the displacement field, which can be expressed as

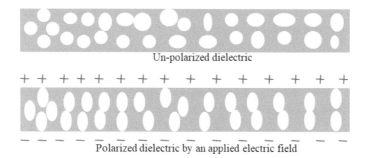

FIGURE 5.3   Schematic representation of polar and non-polar molecules/atoms orientation in the absence of applied electric field and in the presence of applied electric field as in the set-up of parallel plate capacitor mode.

$$\frac{dD}{dE} = \varepsilon_0 \varepsilon_r + N\alpha \qquad (5.7)$$

Therefore, one can say that the value of dD/dE can be positive or negative depending upon the relative permittivity and atomic/molecular polarizability of the material. If D is directly and linearly proportional to applied E in a material, then dD/dE will always be positive, and this happens in linear dielectric materials. However, there are some materials available in the system that are made up of such types of molecules/species, which have their own random dipole moment, revealing non-linear dependency on the applied electric field E of displacement field D, which are known as non-linear dielectric materials. This means $dD/dE = \varepsilon_0 \varepsilon_r$ may be negative, i.e., permittivity will be negative, which reflects negative capacitance (NC). In general, such a phenomenon could be possible with ferroelectric, ferrielectric, and antiferroelectric materials.

5.3   ORIGINS OF NEGATIVE CAPACITANCE IN FERROELECTRICS

Ferroelectric materials are among the most used materials in devices that reflect permanent spontaneous polarization even in the absence of an external electric field owing to their non-centrosymmetric crystal structure that can be reversed under the influence of a stronger external electric field as compared to their coercive field [38]. Such characteristics in materials are more useful in memory storage devices due to their flipping polarization as a function of the electric field [39]. This tuning in the polarization of such crystal-structured materials is also done through variations in temperature and the application of stress to utilize their features in ultrasonic transducers and infrared detectors [40].

The spontaneous permanent polarization of the material depends on its existing molecular structure. Therefore, permanent polarization in any material is the result of its constituent features and the material which has been made up of a set of molecules which contains one or more stable molecules. These molecules change their constituent atom positions with time due to any external forces applied on them. For example, a unit cell (as a representative) of ferroelectric material is shown in Figure 5.4(a), where the central atom (blue) is situated in the upper half (i) of the unit cell and as the applied potential/voltage increases on the ferroelectric material, the central blue atom starts moving to the lower half portion (ii) of the unit cell and comes to a stable position in the lower half part of the unit cell as shown in Figure 5.5(b). As the applied potential/voltage starts to decrease across the ferroelectric it starts to move to its original/initial position, i.e., it shifts to the upper half of the unit cell as shown in Figure 5.5(a). This complete position transition phenomenon in the unit cell defines the material polarization direction or flipping polarization with an external electric field. This complete phenomenon of polarization as a function of an external electric field can be represented as a hysteresis loop, as shown in Figure 5.5(c), in which the blue atom position in the upper half is represented in the first quadrant while the reverse position i.e., the blue atom lower half position, is represented in the third quadrant. The second and fourth quadrant

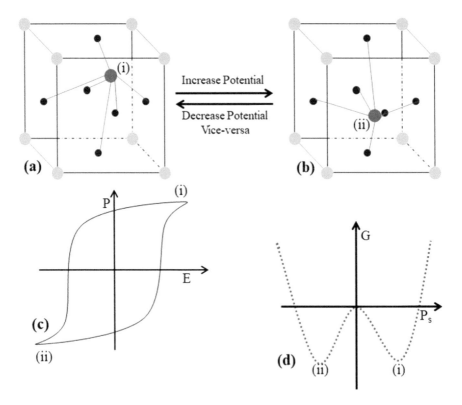

FIGURE 5.4 Schematic representation of atom motion within a ferroelectric unit cell/compound with a change in applied voltage across it (a, b). (c) The polarization versus applied electric field characteristic curve, i.e., hysteresis loop of ferroelectric material and (d) a free energy-polarization plot that reflects two stable positions of the ferroelectric material.

positions are very sharply changed as compared to the first and third quadrants with an electric field, i.e., it is the position of instability of the blue atom in the unit cell. This position of instability is also reflected in the stored energy versus polarization plot [Figure 5.4(d)]. This unstable blue atom position between positions (i) and (ii) is represented in Figure 5.5 within the energy range, i.e., between the stable position, an unstable position of the atom exists, as shown by the half inverted parabola shape. The unstable position energy range starts from zero to nearly half of the stable position energy with respect to positive and negative polarization within the external critical electric field, as shown between the dashed line that represents the negative capacitance region [Figures 5.5(a,b)]. The E_r^{-1} versus P_s plot represents the unstable regions, i.e., (iii) in Figure 5.5(c).

Further, the polarization is a result of dipole moments that are influenced by the permittivity and electric field of the ferroelectric materials, due to which ferroelectric materials are used in microwave electronics so widely [40]. There could be a probable condition during which it reflects negative permittivity in the process of polar transition from one direction to another under the influence of an external electric field. These materials reveal linear as well as non-linear dielectric responses.

One can neglect the linear dielectric (means $P = P$) response to estimate qualitative behavior in the presence of an electric field E_f, and then Equation (5.6) can be approximated as

$$\varepsilon_o \varepsilon = \frac{dD}{dE_f} = \varepsilon_o + \frac{dP_s}{dE_f} \approx \frac{dP_s}{dE_f} \qquad (5.8)$$

Therefore, in general, we can say that when polarization changes its direction under an electric field, its permittivity will be negative and, thus, resultant capacitance will be negative. The first prediction of possible negative capacitance in ferroelectric materials was done by Landauer in 1976 on the basis of the Landau model [6]. This model considers thermodynamic Gibbs free energy (G_f) to evaluate ferroelectric behavior, which was expressed as

$$G_f = \alpha P_s^2 + \beta P_s^4 - E_f P_s \qquad (5.9)$$

where, α and β are Landau coefficients that have negative (i.e., $\alpha < 0$) and positive (i.e., $\beta > 0$) values below Curie temperature. Now, for visualization of Equation (5.8), consider a case when $E_f = 0$ and only second-order phase transition is taking place in G_f, then Gibbs free energy reveals two minima (Figure 5.5d). One minima exists in the negative polarization region and another exists in the positive polarization while in between there is a maxima at zero polarization (i.e., at $P_s = 0$).

These Gibbs free energy minima and maxima are attributed to stable and unstable polarization states of the ferroelectric material since, for the stable polarization state, the Gibbs free energy should be a minimum.

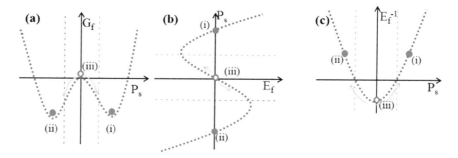

FIGURE 5.5 Schematic curve of (a) free energy-polarization landscape of a ferroelectric material at zero applied voltage, (b) the polarization-electric field dependence (outcome of a), and (c) inverse of permittivity-polarization influence. The solid red ball at positions (i) and (ii) reflect the stable and hollow red ball positions and for unstable condition/position of the ferroelectric material/system.

$$\frac{dG_f}{dP_s} = 0$$

i.e., $\quad E_f = 2\alpha P_s + 4\beta P_s^3$ \hfill (5.10)

This reflects an 'S'-shaped graph between E_f vs. P_s. This plot shows the dP_s/dE_f negative presence in the material for certain values of E_f which is due solely to the permittivity of the ferroelectric material. Now from Equation (5.9), dP_s/dE_f can be written as

$$\frac{dP_s}{dE_f} = \frac{1}{2\alpha + 12\beta P_s^2}$$ \hfill (5.11)

Equations (5.7) and (5.10) illustrate that permittivity is dependent on the α, β, and P_s values. It is clear that for smaller values of P_s the permittivity must be negative owing to the negative slope of the graph in figure, and since α is always negative. This description of negative permittivity (i.e., negative capacitance) was based on the thermodynamical parameters which express the unstable negative permittivity corresponding to maximum Gibbs free energy. It is now possible to implement this in the devices, as explained below.

There are two possible ways to observe the negative permittivity inside a ferro-electric material. First, one can observe it during the switching of the polarization state from positive to negative or vice versa for a limited time interval known as the transient negative capacitance. Second, one can sandwich the ferroelectric material inside a larger sized material such that the negative permittivity state becomes thermo-dynamically stable and overall Gibbs free energy stays at the minimum.

5.3.1 Transient Negative Capacitance

To understand this, let's consider an ideal metal-ferroelectric-metal parallel plate cap-acitor where polarization changes from negative to positive as a function of the applied voltage (Figure 5.6a). The different polarization state positions can be represented by ferroelectric energy landscape sketch diagrams at various applied potential values, such as $V = 0$, $0 < V < V_c$, $V = V_c$, $V > V_c$, $0 < V < V_c$, and $V = 0$, etc. (Figures 5.6b–f). Initially, in the absence of an external potential (i.e., $V = 0$), the energy landscape is symmetric about Gibbs energy (G_f) and the polarization is negative, which represents the stable state.

A small positive potential application lifts the left side minimum energy up to barrier (Figure 5.6c). A further increase in the positive applied potential will lead to almost maximum tilt of the energy landscape toward the right side, except for a particular potential at which $P_s = 0$ for an unstable state of polarization (Figure 5.6d) and as the polarization shifts in the positive region there is an increase in potential (Figure 5.6e). In the reverse of this process as the potential decreases the energy landscape is lifted up on the right side and as the applied potential approaches zero,

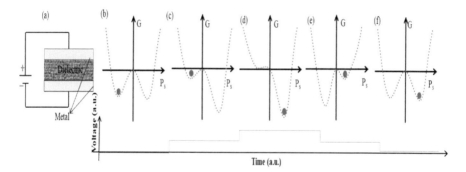

FIGURE 5.6   (a) Schematic diagram of flipping/switching states of a metal-dielectric-metal capacitor system. The free energy-polarization landscape (b) at zero, (c) between zero and the coercive voltage, (d) at higher than coercive voltage, (e) between coercive voltage to zero, and (f) at zero applied voltage.

the energy landscape reach its most stable state in positive polarization due to re-emergence of the energy barrier (Figure 5.6f).

Further, during complete switching, the atom takes some time. This time is an important parameter to express the positions of polarization; the polarization will increase with time ($dP_s/dt' > 0$) and at the same instant the applied voltage/field decreases with time ($dE_f/dt' < 0$). Therefore, in the transient NC case, the measurement prospective can be achieved. To understand this, let's consider a ferroelectric material with free charge (Q) per unit area on the electrode identical to the displacement field, that can be mathematically written as

$$Q = \varepsilon_0 E_f + P_s \tag{5.12}$$

Q estimates the total spontaneous polarization in the ferroelectric. Further, Q varies as a function of time. Therefore, the equation can be represented in the form of a time derivative, as:

$$\frac{dQ}{dt} = \frac{\varepsilon_0 E_f}{dt} + \frac{dP_s}{dt}$$

$$\frac{dE_f}{dt} = \frac{1}{\varepsilon_0}\left(\frac{dQ}{dt} - \frac{dP_s}{dt}\right) \tag{5.13}$$

Since the rate of change of the electric field of a ferroelectric material is dependent on the rate of change of charge and the rate of change of spontaneous polarization, a change in the rate of the electric field defines the material behavior, i.e., it reflects stable or unstable states. Here, for any ferroelectric material, if the rate of change of spontaneous polarization is faster than the rate of change of charge Q, then the rate of change of the electric field will be negative, which leads to a negative capacitance transient situation in the ferroelectric material [8,41–43]. In other words, the rate of transient of electric field explains the switching mechanism and phases of the ferroelectric materials at a microscopic level [41,43]. However, owing to instability in the negative capacitance region, a metal-ferroelectric-metal capacitor will always lead to positive capacitance in any circumstances. Therefore, there is a way to observe the negative capacitance effect/behavior of the ferroelectric material for different purposes. For that, the metal-ferroelectric-metal capacitor has to connect with second circuit elements such as the resistor, dielectric capacitor, or gate of the MOSFET that slow down the rate of change of charge Q and enable the decrease in the ferroelectric voltage with time, in series, to force the voltage across it [44–46]. However, recently Pintilie et al. observed negative capacitance behavior in the ferroelectric material by first increasing the applied voltage followed by a subsequent decrease in the applied voltage with ferroelectric switching [47].

5.3.2 Stabilized Negative Capacitance

Initially, it was agreed by the scientific community that it is almost impossible to observe negative capacitance behavior in a material with stability. However, Salahuddin and Datta suggested in 2008 that the ferroelectric negative capacitance state can be stabilized at zero spontaneous polarization by the addition of an additional layer of material in a series that has positive capacitance [8]. It has also been found that the series connection between the metal-ferroelectric-metal capacitor and the metal-dielectric-metal capacitor leads to large leakage of current as well as the formation of domains that destabilize the negative capacitance behavior of the resultant system [48,49]. Further, when the ferroelectric and dielectric layers of the material are connected in series, i.e., deposited layer by layer, this leads to positive resultant total capacitance with no polarization hysteresis loop [8]. Therefore, for application purposes, the layer-by-layer deposition of ferroelectric and dielectric is the preferred structure for negative capacitance behavior observation [50–55]. To understand the science behind this layer-by-layer preferred structure we have to understand the resultant Gibbs free energy of the total system. The Gibbs free energy of the dielectric material (G_d) can be expressed as:

$$G_d = \frac{D^2}{2\varepsilon_0\varepsilon_d} - E_d D \qquad (5.14)$$

where D, ε_d, and E_d are the displacement field, relative permittivity of the dielectric, and electric field in the dielectric material, respectively. Consider an approximation that $D = P_s$, then the total free energy per unit area of the ferroelectric and dielectric layer-by-layer structure sandwich between two metal plates or electrodes can be expressed with the help of Equations (5.9) and (5.14):

$$G_t = G_f t_f + G_d t_d$$

$$G_t = \left(\alpha t_f + \frac{t_d}{2\varepsilon_0\varepsilon_d} \right) P_s^2 + t_f \beta P_s^4 - VP_s \qquad (5.15)$$

where t_f and t_d represent the thickness of the ferroelectric and dielectric layers. Now, for this system's stability the coefficient of P_s^2 should be positive, i.e.,

$$\alpha t_f + \frac{t_d}{2\varepsilon_0\varepsilon_d} \geq 0$$

$$t_f \leq -\frac{t_d}{2\alpha\varepsilon_0\varepsilon_d} \qquad (5.16)$$

This illustrates that the thickness of the ferroelectric material with respect to the dielectric material plays an important role in the total stability of the negative capacitance effect/behavior of the structured system. Further, relative permittivity of the dielectric material is one of the influencing factors for the stable negative capacitance behavior. To understand the complications, let's consider two cases: first, suppose $t_d \leq t_f$, and second, $t_d \geq t_f$. There will be a critical thickness (t_{cf}) of the ferroelectric material corresponding to the dielectric material for which this system is completely stable. For the first case, the ferroelectric-dielectric layered system is placed between two electrodes as a parallel plate capacitor, as shown in Figure 5.7(a). Here, the free energy variation for different applied voltage ranges as a function of time represents various positions of the stable state of the structured system [Figures 5.7(b–g)]. As the applied voltage increases across the plates, the polarization states start to change their positions according to the resultant electric field inside the material. The increase in the applied voltage, which is function of the time, leads to the final free energy of the system shown below the free energy polarization graph. For the starting stage when the applied voltage is zero, the resultant free energy is negative owing to the prominent influence of the ferroelectric spontaneous polarization. This means that, in this case, free energy will be negative [Figure 5.7(b)]. As applied voltage increases between zero and the critical voltage, the free energy reaches zero or is slightly positive with negative polarization [Figure 5.7(c)]. However, at the critical/saturation applied voltage, the free energy again becomes negative with positive polarization [(Figure 5.7(d)] and above the critical voltage it becomes more negative [Figure 5.7(e)]. Further, as the applied voltage starts to decrease, the free energy again reaches zero or low positive, but this time with positive polarization [Figure 5.7(f)] and as it decreases the applied voltage further to zero, the free energy again becomes negative but with positive polarization [Figure 5.7(f)]. Therefore, in this case we can say that the current sandwich system behaves the same as a metal-ferroelectric-metal capacitor owing to the greater influence of spontaneous polarization of the ferroelectric material as compared to dielectric polarization. This happens due to the larger presence of ferroelectric atoms/molecules resulting from the depolarization field in the ferroelectric as compared to dielectric materials.

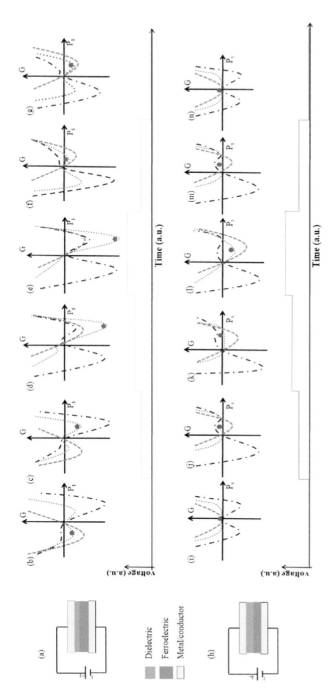

FIGURE 5.7 Schematic diagram of stabilizing negative capacitance in a hybrid metal-dielectric-ferroelectric-metal capacitor configuration on the basis of free energy-polarization landscape. (a) Hybrid system with a smaller size of dielectric material as compared to ferroelectric material. The switching position of system (a) as a function of polarization, (b) at initial zero voltage, (c) between zero and coercive voltage, (d) over coercive voltage, (e) over critical voltage, (f) between coercive voltage to zero voltage, and (g) at coming back to zero voltage. (h) Hybrid system with critical thickness of dielectric material with ferroelectric material (i) at initial zero voltage, (j) between zero and coercive voltage, (k) over coercive voltage, (l) over critical voltage, (m) between coercive voltage and zero voltage, and (n) returning to zero voltage. Voltage as a function of time (yellow and green).

Therefore, the negative capacitance behavior in this system is similar to transient negative capacitance (as discussed in Section 5.4.1).

Now, in the second case, where $t_d \geq t_f$ is considered for the ferroelectric dielectric layer-by-layer structured system fabrication, at the initial stage, i.e., when the applied voltage is zero, the free energy of the fabricated system is zero i.e., the negative capacitance state becomes stable in this system. This happens because the total free energy is minimized due to the addition of the maximum free energy of the ferroelectric material and the minimum free energy of the dielectric material [Figure 5.7(i)]. In other words, we can say that in this situation the free energy and spontaneous polarization of the molecules/atoms present in the ferroelectric material are cancelled out by the constituent molecules/components of the dielectric material. This leads to a very stable negative capacitance effect/behavior. Further, as the positive applied voltages is increased, spontaneous polarization increases, but at the same instant, the electric field decreases owing to depolarization caused by the dielectric material, i.e., dP_s/dE_f is negative, but the system remains stable due to resultant positive free energy [Figure 5.7(j)]. Therefore, total capacitance is always positive. As positive applied voltage increases further, to saturation polarization, the free energy of the fabricated system remains positive, resulting in a stable state with positive capacitance and maximum polarization conditions in the material [Figure 5.7(k)]. However, when the applied voltage increases more than the critical applied voltage, the ferroelectric material enters the positive capacitance state from a stable negative capacitance state [Figure 5.7(l)]. However, as the applied voltage decreases below critical voltage, the ferroelectric material again achieves a negative capacitance state with positive free energy, i.e., with stability [Figure 5.7(m)] and with a further decrease in the applied voltage to zero, the fabricated system again arrives at the initial situation where the free energy is minimum and zero or positive [Figure 5.7(n)]. The variation in the applied voltage as a function of time is grammatically represented below the corresponding free energy polarization plot of the fabricated system. There is an absence of negative free energy except above the critical applied voltage case, i.e., in the whole operating voltage range between zero to critical voltage, the fabricated system is stable with negative capacitance behavior without hysteresis. This fabricated system reflects the best fabrication system with negative capacitance behavior.

It is known that ferroelectric materials are made up of domains that contain their own polarization direction within the material. The size and orientation of these domains define the ferroelectric properties of the material. It contains a large number of atoms/molecules arranged in a set way. If these groups of atoms/molecules, i.e., domains of ferroelectric materials, change their polarization, then they need a large applied voltage to increase the resultant ferroelectric polarization and vice versa. These groups of atoms/molecules, i.e., domains, will also develop prominent depolarization effects. In other words, these domains reduce the depolarization energy of the system.

This means that if we consider a ferroelectric material layer made up of domains over the dielectric layer in the layered structure, then the depolarization energy will reduce faster but in a discontinuous way. However, this type of hybrid ferroelectric-dielectric layered system could emerge with a stable negative capacitance behavior/

state [56]. Further, ferroelectric materials that have homogeneous polarization and that are in a combined layer with dielectric reflect stable and positive free energy, known as intrinsic negative capacitance (INC), while the ferroelectric materials with domains are known to have extrinsic negative capacitance (ENC).

In the simplest case of ferroelectrics with domains, the domains are arranged in the ferroelectric material in such a way that the alternative domains' polarization is in the opposite direction [57]. Then, these domains are on the application of external electric fields (in the presence of an applied voltage. In addition, the domains oriented in the same direction as the electric fields, therefore start increasing their size, while the domains in the opposite direction to the electric field start reducing their size (because of the motion of the domain wall) [58]. This occurs due the special feature of materials that varies from material to material or due to the permittivity of the material and an excellent hybrid configurationally fabricated system with dielectrics that can reflect negative capacitance behavior [43,59,60]. In such cases, the thickness of the ferroelectric layer plays an important role in the system fabricated with dielectric materials to stabilize the negative capacitance without hysteresis formation [41,42,61].

5.4 ORIGIN OF NEGATIVE CAPACITANCE IN ANTIFERROELECTRIC

The ferroelectric and antiferroelectric nature of a material depends on the domain's polarization amplitude and direction. Kittel in his model predicted that antiferroelectric materials reflect two of the same hysteresis loops in the polarization–electric field characteristic curve, as shown in the Figure 5.8(a). The stability of such a phenomenon can be explained from a thermodynamics prospective where the Gibbs free energy-polarization landscape shows the features. The characteristic curve reflects that as the electric field increases, polarization of the material also increases to point b, and then at this point it suddenly/sharply increases to point d without following the initial path of increment and reaches the saturation polarization of the material. However, as the electric field starts decreasing, the polarization follows the same path till point c and below this there is a slightly decrease in the electric field as a result of the sharp decrease in the net polarization, where it reaches point a, i.e., it does not follow the same path, so it reflects the hysteresis loop formation. Again, with a decrease in the electric field to zero, the net polarization reaches zero. Further, with the application of a reverse electric field, the same polarization–electric field curve is obtained as with hysteresis loop formation. Moreover, at a zero electric field, the polarization inside the material will be zero which reflects a stable position at O, as shown in Figure 5.8(b). L and M reflect stable polarization corresponding to a particular electric field where small changes in the electric field did not significantly change the polarization with the same situation in a reverse electric field at points L' and M'. These points reflect the stability position/state of the material as reflected in the free energy–polarization curve at $E_f = E'$ and $E_f = -E'$ in Figures 5.8(c,d).

If we divide the polarization region on the basis of the material response, it reflects three parts, the first is non-polar, the second is forbidden, and the third is

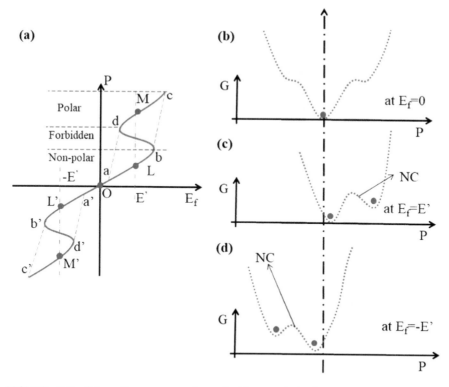

FIGURE 5.8 Schematic representation of (a) a polarization-electric field plot of antiferroelectric material. b'ab corresponds to the region of non-polar antiferroelectric ground state, bd represent the forbidden region, and dc reflects the polar region of polarization of antiferroelectric material under the influence of different electric fields. (b)–(d) The antiferroelectric free energy landscape at different electric fields.

a polar region. In the non-polar region at $E_f = 0$ the material is in a ground state. With a further increase in the electric field to $E_f = E$ in the non-polar region, polarization changes continuously with an electric field, whereas in the forbidden region, it changes abruptly, that reflects basically the negative capacitance behavior of the antiferroelectric material, where d^2G/dP^2 is less than zero. Moreover, the free energy of the polar and non-polar regions depends on the antiferroelectric material and strength of the applied electric field. Therefore, the polar state/region free energy could be lower as compared to the non-polar region, as a result of the negative capacitance at the antiferroelectric transition in a stabilized microscopic system [6,8]. This type of stable microscopic system can be fabricated with a hybrid configuration of antiferroelectric materials with suitable dielectric materials where positive capacitance of the dielectric can reduce or minimize the effect of negative capacitance of the antiferroelectric [61–63]. This could result in negative capacitance in the hybrid system with overall positive capacitance, i.e., with a stable system/structure for applications [8,64].

5.5 STOICHIOMETRIC CONFIGURATION ROLE IN NEGATIVE CAPACITANCE

The hybrid metal-dielectric-ferroelectric-metal capacitor system or metal-dielectric-antiferroelectric-metal capacitor structured system is fabricated by unique techniques such as atomic layer deposition, pulsed layer deposition, etc., where layers of different thicknesses and configurations are deposited onto one another depending upon the requirement. This type of hybrid system can be generated by an ion implantation technique due to the targeted phase, and structural and stoichiometric variations in the material up to the desired penetration depth. Doping of the desirable component during the fabrication process also reflects hybrid system fabrication [65]. In this way we can say that the negative capacitance behavior of the hybrid system not only depends on the thickness of the layer thickness but also their chemical composition.

Li et al., studied a MOSFET system with its ferroelectric component to observe the thickness effect on the negative capacitance behavior of a device, and observed that as the thickness of the ferroelectric layer is increased the subthreshold slope of the device decreases due to the negative capacitance feature of ferroelectric material in the device (Table 5.1), however, in order to minimize the size of the device one can't keep increasing the thickness of the ferroelectric layer [66]. Therefore, FE thickness should be properly selected in order to achieve both current enhancement and saturated output characteristics.

Similarly, we know that temperature is the one of the most important factors in device life and performance. Xiao et al. reported a study of NCFETs as a function of temperature in a large temperature range of 280–360 K (Table 5.2) [67]. They claimed that as temperature increases the device power consumption increases, with poor performance of the transistor device. Therefore, according to their study, lower operating temperature gives better device performance as compared to higher operating temperatures.

TABLE 5.1
A comparison of various parameters

S. no.	T_{FE} (nm)	SS (mV/decade)	Remarks
1	0	88	
2	40	82	↑ ↓
3	80	75	
4	120	68	

T_{FE}(nm)   SS (mV/decade)

TABLE 5.2
The effect of temperature and subthreshold swing

S.N.	Temperature (K)	SS (mV/decade)	Remarks
1	280	56.5	
2	**300**	**59.0< 60.0**	
3	320	64.7	
4	340	68.2	
5	360	71.2	

Temperature   SS

Adapted from Xiao, Y. G., Tang, M. H., Li, J. C., Cheng, C. P., Jiang, B., Cai, H. Q., ... & Gu, X. C. (2012). Temperature effect on electrical characteristics of negative capacitance ferroelectric field-effect transistors. *Applied Physics Letters* 100(8): 083508.

5.6   CONCLUSION

The current CMOS technology cannot fulfill today's global requirements owing to the non-scalability of threshold voltage due to immutable subthreshold swing (SS) leading to higher leakage current for low-power circuit operation. This has necessitated the search for novel low threshold voltage or steep-slope devices such as tunnel field effect transistors (TFETs) negative capacitance FETs (NC-FETs), positive feedback FETs, and impact ionization MOSFETs (IMOSs) that can replace or complement the existing CMOS technology. Negative capacitance is a newly discovered state of ferroelectric materials that enhances the electronic application of ferroelectric materials along with high dielectric materials with appropriate hybrid structure configuration. This NCFET behavior/state could reduce the SS, size, and also power dissipation of devices. The physics and material engineering behind the NC state enable selection of the most suitable composition of ferroelectric and dielectric materials, which is crucial for excellent and optimized performance. Antiferroelectric materials also exhibit negative capacitance behavior due to their hysteresis features. Their proper combination with dielectric material leads to stable negative capacitance behavior/state due to minimization of the free energy of the whole system. Negative free energy is countered by dielectric free energy at a particular polarization state or at a particular time of the voltage swing. In microscopically designed structural devices, the roles of the concentration of parent and dopant molecules, domain size, domain wall movement, size of the layered structure, and applied voltage with time should be optimized to achieve the required performance of negative capacitance behavior of the materials. The proposed devices have revealed excellent features that will reduce global issues with increased operating speed of devices.

REFERENCES

1. K. N. Tu (2003) Recent advances on electromigration in very-large-scale-integration of interconnects. *Journal of Applied Physics* 94: 5451.
2. S. B. Rahi, S. Tayal, A. Kumar (2021) A review on emerging negative capacitance field effect transistor for low power electronics. *Microelectronics Journal* 116: 1052.
3. Tayal, Shubham, Abhishek Kumar Upadhyay, Deepak Kumar, and Shiromani Balmukund Rahi (eds.) (2022) *Emerging Low-Power Semiconductor Devices: Applications for Future Technology Nodes*. CRC Press.
4. A. K. Upadhyay, S. B. Rahi, S. Tayal, Y. S. Song (2022) Recent progress on negative capacitance tunnel FET for low-power applications: Device perspective. *Microelectronics Journal* 129: 105583.
5. S. Tayal, S. B. Rahi, J. P. Srivastava, S. Bhattacharya (2022) Recent trends in compact modeling of negative capacitance field effect transistors. In: *Semiconductor Devices and Technologies for Future Ultra-low Power Electronics*. CRC.
6. R. Landauer (1976) Can capacitance be negative? *Collective Phenomena* 2: 167–170.
7. Y. Hong and C. Shin (2020) Yield estimation of NCFET-based 6-T SRAM. In: *2020 4th IEEE Electron Devices Technology & Manufacturing Conference (EDTM)*, pp. 1–3.
8. S. Salahuddin and S. Datta (2008) Use of negative capacitance to provide voltage amplification for low power nanoscale devices. Nano Letters 8: 405–410.
9. Michael Hoffmann, Stefan Slesazeck and Thomas Mikolajick (2021) Progress and future prospects of negative capacitance electronics: A materials perspective. *APL Materials* 9: 020902.
10. Michael Hoffmann, Zheng Wang, Nujhat Tasneem, Ahmad Zubair, Prasanna Venkatesan Ravindran, Mengkun Tian, Anthony Arthur Gaskell, Dina Triyoso, Steven Consiglio, Kandabara Tapily, Robert Clark, Jae Hur, Sai Surya Kiran Pentapati, Sung Kyu Lim, Milan Dopita, Shimeng Yu, Winston Chern, Josh Kacher, Sebastian E. Reyes-Lillo, Dimitri Antoniadis, Jayakanth Ravichandran, Stefan Slesazeck, Thomas Mikolajick & Asif Islam Khan (2022) Antiferroelectric negative capacitance from a structural phase transition in zirconia. *Nature Communications* 13: 1228.
11. C. Auth et al. (2017) A 10nm high performance and low-power CMOS technology featuring 3rd generation FinFET transistors, self-aligned quad patterning, contact over active gate and cobalt local interconnects. In: *IEEE International Electron Devices Meeting (IEDM)* (IEEE, San Francisco, CA, USA), pp. 29.1.1–29.1.4.
12. L.-A. Ragnarsson, T. Chiarella, M. Togo, T. Schram, P. Absil, and T. Hoffmann (2011) Ultrathin EOT high-κ/metal gate devices for future technologies: Challenges, achievements and perspectives. *Microelectronic Engineering* 88: 1317–1322.
13. V. V. Zhirnov and R. K. Cavin (2008) Negative capacitance to the rescue? *Nature Nanotechnology* 3: 77–78.
14. S. M. Sze and K. K. Ng (2007) *Physics of Semiconductor Devices*. John Wiley & Sons, Chichester, UK.
15. Y. Hong, Y. Choi and C. Shin (2020) NCFET-based 6-T SRAM: Yield estimation based on variation-aware sensitivity. *IEEE Journal of the Electron Devices Society* 8: 182–188.
16. K. Lee et al. (2019) Analysis on fully depleted negative capacitance field-effect transistor (NCFET) based on electrostatic potential difference. In: *2019 Electron Devices Technology and Manufacturing Conference (EDTM)*, pp. 422–424.
17. X. Sun, Y. Zhang, J. Xiang, K. Han, X. Wang and W. Wang (2021) Role of interfacial traps at SiO_2/Si interface in negative capacitance field effect transistor (NCFET) based

on transient negative capacitance (NC) theory. In: *2021 5th IEEE Electron Devices Technology & Manufacturing Conference (EDTM)*, pp. 1–3.

18. S. Roy, P. Chakrabarty and R. Paily (2022) Assessing RF/AC performance and linearity analysis of NCFET in CMOS-compatible thin-body FDSOI. *IEEE Transactions on Electron Devices* 69(2): 475–481.

19. H. Wang et al. (2018) New insights into the physical origin of negative capacitance and hysteresis in NCFETs. In: *2018 IEEE International Electron Devices Meeting (IEDM)*, pp. 31.1.1–31.1.4.

20. D. Kwon, S. Cheema, N. Shanker, K. Chatterjee, Y.-H. Liao, A. J. Tan, C. Hu, and S. Salahuddin (2019) Negative capacitance FET with 1.8-nm-thick Zr-doped HfO_2 oxide. *IEEE Electron Device Letters* 40: 993–996.

21. D. Kwon, S. Cheema, Y.-K. Lin, Y.-H. Liao, K. Chatterjee, A. J. Tan, C. Hu, and S. Salahuddin (2020) Near threshold capacitance matching in a negative capacitance FET with 1nm effective oxide thickness gate stack. *IEEE Electron Device Letters* 41: 179–182.

22. N. Liu et al. (2022) Reconfigurable ferroelectric electrostatic doped negative capacitance nanosheet field-effect transistors with enhanced ION/IOFF and scaled VDD < 0.45 V. In: *2022 6th IEEE Electron Devices Technology & Manufacturing Conference (EDTM)*, pp. 288–290.

23. X. Huang et al. (2021) A dynamic current model for MFIS negative capacitance transistors. *IEEE Transactions on Electron Devices* 68(7): 3665–3671.

24. Z. C. Yuan et al. (2019) Toward microwave S- and X-parameter approaches for the characterization of ferroelectrics for applications in FeFETs and NCFETs. *IEEE Transactions on Electron Devices* 66(4): 2028–2035.

25. C.-C. Fan, C.-H. Cheng, Y.-R. Chen, C. Liu and C.-Y. Chang (2017) Energy-efficient HfAlOx NCFET: Using gate strain and defect passivation to realize nearly hysteresis-free sub-25mV/dec switch with ultralow leakage. In: *2017 IEEE International Electron Devices Meeting (IEDM)*, pp. 23.2.1–23.2.4.

26. T. Cam et al. (2020) Sustained benefits of NCFETs under extreme scaling to the end of the IRDS. *IEEE Transactions on Electron Devices* 67(9): 3843–3851.

27. J. K. Wang et al. (2022) Potential enhancement of fT and g_mfT/ID via the use of NCFETs to mitigate the impact of extrinsic parasitics. *IEEE Transactions on Electron Devices* 69(8): 4153–4161.

28. G.-Y. He et al. (2021) Simultaneous analysis of multi-variables effect on the performance of multi-domain MFIS negative capacitance field-effect transistors. *IEEE Journal of the Electron Devices Society* 9: 741–747.

29. H. Liu et al. (2021) Analysis of using negative capacitance FETs to optimize linearity performance for voltage reference generators. *IEEE Transactions on Electron Devices* 68(11): 5864–5871.

30. G. Yang et al. (2020) Scaling MoS_2 NCFET to 83 nm with record-low ratio of SSave/SSRef.=0.177 and minimum 20 mV hysteresis. In: *2020 IEEE International Electron Devices Meeting (IEDM)*, pp. 12.4.1–12.4.4.

31. R. A. Vega, T. Ando and T. M. Philip (2021) Junction design and complementary capacitance matching for NCFET CMOS logic. *IEEE Journal of the Electron Devices Society* 9: 691–703.

32. Y. Liang et al. (2018) Influence of body effect on sample-and-hold circuit design using negative capacitance FET. *IEEE Transactions on Electron Devices* 65(9): 3909–3914.

33. Sharma and K. Roy (2017) Design space exploration of hysteresis-free HfZrOx-based negative capacitance FETs. *IEEE Electron Device Letters* 38(8): 1165–1167.

34. S. Luo, X. Zhang and G. Liang (2020) Performance evaluation and device physics investigation of negative-capacitance MOSFETs based on ultrathin body silicon and monolayer MoS_2. *IEEE Transactions on Electron Devices* 67(8): 3049–3055.
35. W.-X. You and P. Su (2017) Design space exploration considering back-gate biasing effects for 2D negative-capacitance field-effect transistors. *IEEE Transactions on Electron Devices* 64(8): 3476–3481.
36. R. C. Bheemana, A. Japa, S. Yellampalli and R. Vaddi (2021) Steep switching NCFET based logic for future energy efficient electronics. In: *2021 IEEE International Symposium on Smart Electronic Systems (iSES)*, pp. 327–330.
37. Y. Hong and C. Shin (2020) Yield estimation of NCFET-based 6-T SRAM. In: *2020 4th IEEE Electron Devices Technology & Manufacturing Conference (EDTM)*, pp. 1–3.
38. M. E. Lines and A. M. Glass (1977) *Principles and Applications of Ferroelectrics and Related Materials*. Oxford University Press.
39. T. Mikolajick, U. Schroeder, and S. Slesazeck (2020) The past, the present, and the future of ferroelectric memories. *IEEE Transactions on Electron Devices* 67: 1434–1443.
40. A. K. Tagantsev, V. O. Sherman, K. F. Astafiev, J. Venkatesh, and N. Setter (2003) Ferroelectric materials for microwave tunable applications. *Journal of Electroceramics* 11: 5–66.
41. C.-I. Lin, A. I. Khan, S. Salahuddin and C. Hu (2016) Effects of the variation of ferroelectric properties on negative capacitance FET characteristics. *IEEE Transactions on Electron Devices* 63(5): 2197–2199.
42. Z. C. Yuan et al. (2016) Switching-speed limitations of ferroelectric negative-capacitance FETs. *IEEE Transactions on Electron Devices* 63(10): 4046–4052.
43. Z. Zheng et al. (2018) Real-time polarization switch characterization of $HfZrO_4$ for negative capacitance field-effect transistor applications. *IEEE Electron Device Letters* 39(9): 1469–1472.
44. A. I. Khan, K. Chatterjee, B. Wang et al. (2015) Negative capacitance in a ferroelectric capacitor. *Nature Materials* 14: 182–186.
45. A. I. Khan, M. Hoffmann, K. Chatterjee, Z. Lu, R. Xu, C. Serrao, S. Smith, L. W. Martin, C. Hu, R. Ramesh, and S. Salahuddin (2017) Differential voltage amplification from ferroelectric negative capacitance. *Applied Physics Letters* 111: 253501.
46. A. I. Khan et al. (2016) Negative capacitance in short-channel FinFETs externally connected to an epitaxial ferroelectric capacitor. *IEEE Electron Device Letters* 37: 111–114.
47. L. Pintilie, G. A. Boni, C. Chirila, L. Hrib, L. Trupina, L. D. Filip, and I. Pintilie (2020) Polarization switching and negative capacitance in epitaxial PbZr0.2Ti0.8O3 thin films. *Physical Review Applied* 14: 014080.
48. M. Hoffmann, M. Pešiʹc, S. Slesazeck, U. Schroeder, and T. Mikolajick (2018) On the stabilization of ferroelectric negative capacitance in nanoscale devices. *Nanoscale* 10: 10891–10899.
49. A. I. Khan, U. Radhakrishna, K. Chatterjee, S. Salahuddin, and D. A. Antoniadis (2016) Negative capacitance behavior in a leaky ferroelectric. *IEEE Transactions on Electron Devices* 63: 4416–4422.
50. S. Salamin, G. Zervakis, Y. S. Chauhan, J. Henkel and H. Amrouch (2021) PROTON: Post-synthesis ferroelectric thickness optimization for NCFET circuits. *IEEE Transactions on Circuits and Systems I: Regular Papers* 68(10): 4299–4309.
51. S. Kim, K. Lee, J.-H. Lee, B.-G. Park and D. Kwon (2021) Gate-first negative capacitance field-effect transistor with self-aligned nickel-silicide source and drain. *IEEE Transactions on Electron Devices* 68(9): 4754–4757.

52. Z. C. Yuan, P. S. Gudem, A. Aggarwal, C. VanEssen, D. Kienle and M. Vaidyanathan (2021) Feedback stabilization of a negative-capacitance ferroelectric and its application to improve the fT of a MOSFET. *IEEE Transactions on Electron Devices* 68(10): 5101–5107.

53. R. R. Shaik and K. P. Pradhan (2023) Investigation on impact of doped HfO$_2$ thin film ferro-dielectrics on FDSOI NCFET under back-gate bias influence. *IEEE Transactions on Nanotechnology* 22: 14–19.

54. H. Lee, M. Sritharan and Y. Yoon (2022) A computational framework for gradually switching ferroelectric-based negative capacitance field-effect transistors. *IEEE Transactions on Electron Devices* 69(10): 5928–5933.

55. C.-S. Hsu, C. Pan and A. Naeemi (2018) Performance analysis and enhancement of negative capacitance logic devices based on internally resistive ferroelectrics. *IEEE Electron Device Letters* 39(5): 765–768.

56. P. Zubko, J. C. Wojdeł, M. Hadjimichael, S. Fernandez-Pena, A. Sene, I. Luk'yanchuk, J.-M. Triscone, and J. Iniguez (2016) Negative capacitance in multidomain ferroelectric superlattices. *Nature* 534: 524–528.

57. A. Kopal, T. Bahnik, and J. Fousek (1997) Domain formation in thin ferroelectric films: The role of depolarization energy. *Ferroelectrics* 202: 267–274.

58. A. Kopal, P. Mokry, J. Fousek, and T. Bahnik (1999) Displacements of 180° domain walls in electroded ferroelectric single crystals: The effect of surface layers on restoring force. *Ferroelectrics* 223: 127–134.

59. I. Luk'yanchuk, A. Sene, and V. M. Vinokur (2018) Electrodynamics of ferroelectric films with negative capacitance. *Physical Review B* 98: 024107.

60. H. W. Park, J. Roh, Y. B. Lee, and C. S. Hwang (2019) Modeling of negative capacitance in ferroelectric thin films. *Advanced Materials* 31: 1805266 (2019).

61. J. Iniguez, P. Zubko, I. Luk'yanchuk, and A. Cano (2019) Ferroelectric negative capacitance. *Nature Reviews Materials* 4: 243–256.

62. Kim, Y. J. et al. (2016) Time-dependent negative capacitance effects in Al$_2$O$_3$/BaTiO$_3$ bilayers. *Nano Letters* 16: 4375–4381.

63. Hoffmann, M. et al. (2019) Unveiling the double-well energy landscape in a ferroelectric layer. *Nature* 565: 464–467.

64. Hoffmann, M. et al. (2019) Negative capacitance for electrostatic supercapacitors. Advanced Energy Materials 9: 1901154.

65. C.-H. Cheng et al. (2019) Investigation of gate-stress engineering in negative capacitance FETs using ferroelectric hafnium aluminum oxides. *IEEE Transactions on Electron Devices* 66(2): 1082–1086.

66. Y. Li, Y. Kang and X. Gong (2017) Evaluation of negative capacitance ferroelectric MOSFET for analog circuit applications. *IEEE Transactions on Electron Devices* 64(10): 4317–4321.

67. Xiao, Y. G., Tang, M. H., Li, J. C., Cheng, C. P., Jiang, B., Cai, H. Q., ... & Gu, X. C. (2012) Temperature effect on electrical characteristics of negative capacitance ferroelectric field-effect transistors. *Applied Physics Letters* 100(8): 083508.

6 Basic Operation Principle of Optimized NCFET
Amplification Perspective

S. Yadav[1], P.N. Kondekar[2], B. Awadhiya[3]
[1]Department of Electronics and Communication Engineering, PDPM-Indian Institute of Information Technology, Design & Manufacturing, Jabalpur, M.P., 482005
[2]Department of Electronics and Communication Engineering, PDPM-Indian Institute of Information Technology, Design & Manufacturing, Jabalpur, M.P., 482005
[3]Department of Electronics and Communication, Manipal Institute of Technology, Manipal Academy of Higher Education, Manipal, Udupi, Karnataka, 576104

6.1 INTRODUCTION

As the scaling of CMOS ICs progresses, power leakage/consumption is one of the important issues to address in ultra-low-power technology design. Although scaling power supply voltage is one of the effective ways to reduce dynamic power dissipation, i.e., $P_{dynamic} \propto V_{DD}^2$, this comes at the cost of reducing the ON current of the device, which ultimately compromises the transistor speed. This results in the adjustment of the threshold voltage to maintain the ON current i.e., $I_{ON} = \left(V_{DD} - V_{TH} \right)^\alpha$, where α is between 1 and 2. However, lowering the threshold voltage increases the OFF current of the device, i.e., $I_{OFF} \propto 10^{-V_{TH}/ss}$, which again increases the device static power consumption, $P_{Static} \propto I_{OFF}$. This contradictory requirement implies a trade-off between power and performance at lower technology nodes. However, lowering the device's subthreshold swing (SS) may be one approach to alleviate this issue. The reduction in SS enables simultaneous scaling of V_{DD} and V_{TH} without sacrificing performance or power dissipation. Moreover, the SS of a classical FET, on the other hand, cannot be decreased below the Boltzmann limit (60 mV/decade). The difficulty in lowering SS [1] below the Boltzmann limit is a significant and long-standing problem. As a result, a worldwide search is underway for an ideal switch that can overcome Boltzmann tyranny and offer SS of less than 60 mV per decade. Over a decade, to overcome the fundamental Boltzmann limits and to achieve further scaling in a conventional CMOS technology, steep switches like tunnel-FET [2], phase-FET [3], hybrid-FET [4-6], etc. have received a great deal of attention. The negative capacitance-based field effect transistor has attracted a lot of attention and popularity among the various steep switching devices. A lot of research also has been done in this area due to its CMOS compatibility and easiness of use in complex fabrications. To accomplish the SS < 60 mV/decade, NCFET utilizes the advantage of a differential negative capacitance effect that occurs in ferroelectric materials under some constraints. In this chapter, Section 6.3 places emphasis on the basic notion of negative capacitance's physical origin, stability, and history in attaining steep switching benefits in FETs. The basic operating principle of an NCFET, and its various device architectures, namely MFMIS and MFIS used for low-power applications, will be discussed in Section 6.4. Unique NCFET device features, such as SS, NDIBL, and NDR, which can be useful in ultra-low power steep switching logics and memory applications, are discussed in Section 6.5. The various modeling techniques used for NCFETs are discussed in Section 6.6. The conclusion is in Section 6.7 and the future scope is described in Section 6.8 .

6.2   BASICS OF NEGATIVE CAPACITANCE MATERIALS

These are materials from the family of ferroelectric/antiferroelectric oxides that show negative capacitance properties (under some conditions) on the application of an electric field. Fundamentally, a device that has the capability of storing charge is defined as a capacitor. The capacitance C is formulated as the rate of change of charge (Q) with respect to voltage (V), i.e., $C = dQ/dV$. Capacitance can be positive or negative. If Q increases as V increases it is defined as positive capacitance. On the other hand, if Q decreases as V increases or Q increases as V decreases it is termed negative capacitance. Capacitance can also be expressed as a function of energy U (Equation 6.1). For linear capacitors, capacitance in terms of free energy is given by Equation 6.2.

$$U = \frac{Q^2}{2C} \tag{6.1}$$

$$C = \left[\frac{d^2U}{dQ^2} \right]^{-1} \tag{6.2}$$

A nonlinear capacitor also follows the same relationship. In the case of a positive capacitor, the energy landscape is a parabola, while an inverted parabola occurs for a negative capacitor, as shown in Figures 6.1(a) and (b). An insulating material's energy landscape has a negative curvature zone that corresponds to a negative capacitance. Figure 6.1(c) depicts the ferroelectric material's energy profile. There are two degenerate energy minima in it. This indicates that even without an applied electric field, the ferroelectric material could provide non-zero polarization. The net charge density can be defined as $Q_f = \varepsilon E + P$, where ε is the ferroelectric's linear permittivity, external electric field and polarization are denoted by E and P, respectively. Generally, $P \gg E$ in ferroelectric materials, resulting in $Q_f \approx P$. Comparing the ferroelectric energy profile [Figure 6.1(c)], it can be seen that the curvature of a ferroelectric around $Q = 0$

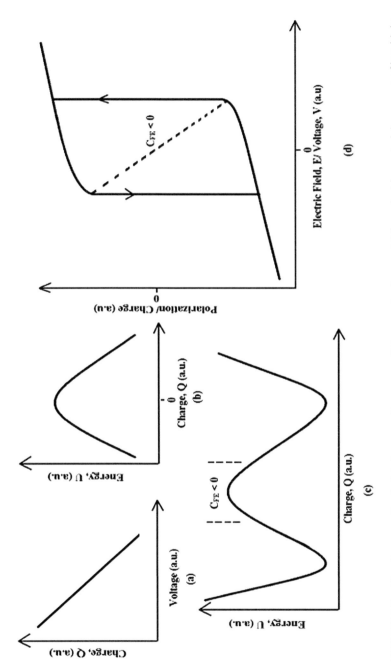

FIGURE 6.1 Negative capacitor in (a) charge–voltage curve, (b) energy landscape, (c) ferroelectrics material energy profile with its negative capacitance region, and (d) its hysteresis charge–voltage/polarization–electric field loop with its S-curve showing negative capacitance region.

is exactly the same as that of a negative capacitor [Figure 6.1(b)]. This curvature suggests that the ferroelectric material has a negative capacitance near $Q = 0$. As a result, a ferroelectric substance could give a negative capacitance around this region. It can also be said that at $E = 0$ ferroelectric materials exhibit non-zero spontaneous polarization and behave as insulating materials having more than one discrete stable or metastable state. A ferroelectric material should switch between these states during $E > E_C$, where E_C is a coercive field. These changes in the energy states occur due to the field polarization coupling [Figure 6.1 (d)].

Joseph Valasek demonstrated the hysteresis characteristics of the polarization of Rochelle salt [7], $NaKC_4H_4O_6:4H_2O$, and its dependence on temperature in 1921, which established the polarization switching-ability factors for ferroelectrics and, in fact, the name 'ferroelectricity.' Then, various ferroelectric ceramic materials like inorganic $BaTiO_3$ [8] and perovskite ($PbZrO_3$, $PbTiO_3$) [9] were discovered during the 1950s to 1970s. However, the discovery of hafnium dioxide (HfO_2) as ferroelectric paved the way for its use with the existing CMOS process. The compatibility of HfO_2 with silicon (Si), high coercive field, thickness scalability, and requiring thinner oxides (≈ 4 nm) to induce ferroelectricity are the key features for its uses in ultra-low-power applications. To date, many doped HfO_2 compounds, for example, doped Sr [10], Al [11,12], La [13], Y [14,15], Gd [16], Zr [17–20], and Si [18,21–25] are used to induce ferroelectricity. Additionally, if HfO_2 is not doped and is deposited using chemical vapor deposition rather than physical vapor deposition, which is dielectric in nature, HfO_2 will exhibit ferroelectricity [26,27].

The non-centro-symmetry occurring in the ferroelectric oxides crystal is the reason behind ferroelectricity. This is necessary for the formation of electric dipoles, and consequently the vector quantity polarization. The off-centering ions create electric polarization or spontaneous dipole moments in the material and correspond to the two separate minimums of the ferroelectric energy landscape. Figure 6.1(d) depicts the energy curve for the corresponding hysteresis P-E loop of ferroelectric [28,29]. This hysteresis loop was well explained by Landau [30] in 1937. The theory of Landau explained the phenomenology model based on symmetry that acts as a conceptual link between microscopic models and observed macroscopic events. Landau emphasized in his writings that in a system smooth transition cannot happen between two phases with distinct symmetries. Because thermodynamically, states of two symmetry-wise different phases should be the same between their shared transition line, one symmetric phase must be greater than the other. Landau subsequently defined these transitions as an order parameter, a physical entry whose value is zero at high symmetry phase and gradually increases to a finite value at the low symmetry phase. This order parameter represents polarization in the case of a ferroelectric–paraelectric transition, and the low and high symmetry phases correspond to the ferroelectric and paraelectric states. Only symmetry-compatible terms are kept when the free energy U is extended as a power series of the order parameter P. To acquire the spontaneous polarization P_0, the system's state is determined by the minimal free energy $U(P)$ with regard to P. Experiments or first-principle calculations can be used to find the coefficients of the series expansion $U(P)$. The $U(P)$ of a ferroelectric is expressed in terms of the even-order polynomial of the polarization P, as given in Equation 6.3.

$$U = \alpha P^2 + \beta P^4 + \gamma P^6 - EP \tag{6.3}$$

The applied electric field is $E = V/d$, and the voltage across the ferroelectric and the ferroelectric thickness are V and d, respectively. The anisotropy constants are α, β, and γ, and β and γ are insensitive to temperature. γ is a positive quantity. For second-order and first-order phase transitions, β is positive and negative, respectively. $\alpha = \alpha_0 (T - T_C)$, where α_0 is a positive and temperature-independent quantity. T and T_C are defined as the temperature and Curie temperature. When α is below the Curie temperature this condition leads to the negative curvature in a ferroelectric's energy-landscape (at $P = 0$). In addition, this condition leads to the double-well energy curve. The application of an electric field tilts the energy profile by $-EP$, which results in a variation of the curve, as shown in Figure 6.1(d). For $T < T_C$, Equations 6.2 and 6.3 result in capacitance equations for $P \oplus 0$ (Equation 6.4). Also, for $dU/dP = 0$, the electric field can be given by Equation 6.5 and P–V curves using this are shown in Figure 6.1(d).

$$C = \frac{1}{2\alpha_0 (T - T_C)} < 0 \tag{6.4}$$

$$E = 2\alpha P + 4\beta P^3 + 6\gamma P^5 \tag{6.5}$$

It should be noted that the ferroelectric-based capacitor shows certain regions (dashed curve) where negative capacitance can be obtained due to nonlinear Q–V characteristics. Landau theory is extensively used to study negative capacitance but due to the unstable nature of negative capacitance in ferroelectric experimental measurement, this was not possible during the early years. However, some groups [31,32] later showed a negative capacitance state in materials, providing experimental evidence of its use in low-power applications.

6.3 NCFET OVERVIEW

In order to replace or improve state-of-the-art transistors, negative capacitance-based FETs have attracted a lot of attention from several steep switching devices. In reality, understanding how NCFETs work begins with a simple SS equation (Equations 6.6, 6.7) which is composed of various capacitances in the FET.

$$SS = \frac{KT}{q} \ln 10 \frac{\partial V_G}{\partial \varphi_s} \tag{6.6}$$

$$\frac{\partial V_G}{\partial \varphi_s} = m = 1 + \frac{C_S}{C_{INS}} \tag{6.7}$$

Insulating/oxide and channel surface region capacitance are denoted by C_{INS} and C_S. The negative capacitance effect allows the 'm' factor (which was previously assumed to be impossible by having a value lower than '1') to be lower than '1' in these equations. This would be the most distinguishing feature of NCFETs which allows SS to be lower than 60 mV/decade.

NCFET research came into the limelight when Salahuddin et al. [33] provided the concept of utilizing the negative capacitance property in ferroelectrics for achieving ultra-low-power and sub-60 mV/decade SS. They suggested that utilizing a hetero-structure of ferroelectric and gate oxide in FET voltage amplification could be achieved, which helps in overcoming the fundamental limit (Boltzmann tyranny) in conventional FinFET/MOSFET. Later research has shown that in order to stabilize the ferroelectric material's inherent unstable negative capacitance, a ferroelectric layer must be deposited on top of the gate oxide layer. Initially, they used PZT $[Pb(Zr_xTi_{1-x})O_3]$ as the ferroelectric material. As discussed in Section 6.3, in terms of energy, ferroelectric materials have two polarized stable states. These states switch from one state to the other under an externally applied bias. When increasing applied bias reaches the coercive voltage/electric field in ferroelectric, its polarization state switches. This switching results in negative differential charge–voltage variations ($dQ/dV < 0$) and capacitance associated with it as negative capacitance. This results in the ferroelectric-gate oxide hetero-structure C_{INS} being negative, which makes $m < 1$ in the SS equation (Equation 6.5) and achieves SS < 60 mV/decade at 300 K, which is shown in the general transfer characteristics (Figure 6.2).

After introduction of the NCFET concept, a few trailblazing experiments were carried out to empirically determine the existence of negative capacitance [34,35]. Khan et al. used a simple RC circuit; where PZT with 60 nm thickness acts as a capacitor in series with a resistor to directly measure negative capacitance [28]. With the help of the observed voltage–time plot, they demonstrated a negative slope, which confirms the negative differential capacitance's existence. In 2014, Appleby et al. and Zhao et al.

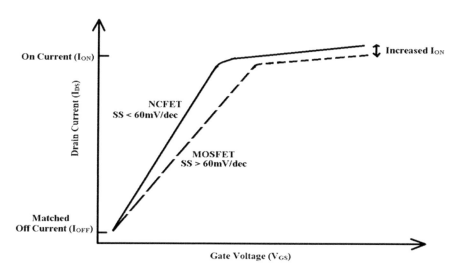

FIGURE 6.2 NCFET transfer curve showing NCFET achieving a steep subthreshold swing compared with MOSFET.

[36,37] experimentally verified that in a series-connected hetero-structure of $SrTiO_3$ and $BaTiO_3$ capacitance increases with $BaTiO_3$ thickness, which confirms the role of BTO's negative capacitance in overall capacitances. In the same year, Gao et al. [38] provided similar evidence of negative capacitance by using $Ba_{0.8}Sr_{0.2}TiO_3 + LaAlO_3$ hetero-structure. Similarly, Ku et al. [39] and Sharma et al. [43] have also confirmed the negative capacitance phenomenon in $P(VDF_{0.75}-TrFE_{0.25})$ and $Hf_{0.5}Zr_{0.5}O_2$ (HZO), respectively. Jo et al. performed a proof-of-concept work in 2015, demonstrating the sub-60-mV/dec. advantage of NCFETs by linking in series, the conventional MOSFET gate terminal and ferroelectric capacitor [40] and achieving a subthreshold slope of 18 mV/decade at 300 K. In 2016, Khan et al. used the FinFET with a ferroelectric [$BiFeO_3$ (BFO)] gate stack to form NCFinFET. Many articles dealing with negative capacitance MOSFETs [41–46] and FinFETs [47–50] demonstrate the reproducible NC effect. Some recent works have theoretically and experimentally demonstrated the advantages of utilizing negative capacitance using state-of-the-art devices such as 2-D FETs [51–54], nanoelectromechanical switches [55,56], carbon nanotubes [57], SOI devices [58], and polymer-ferroelectric FETs [59]. It is evident that for any logic application transistor's threshold voltage should be equal during turning on and off in a device. However, the ferroelectrics employed in NCFETs have a unique 'hysteresis' property that prevents NCFETs from being used as logic transistors [60]. Therefore, obtaining a hysteresis-free operation capacitor matching in dielectric + ferroelectric hetero-structure is required. Ferroelectric thickness optimization is done to achieve hysteresis-free operation but this increases the SS of the transistor [41,50,61]. Therefore, the trade-off between the SS and hysteresis window is important to achieve performance benefits in NCFETs. Ferroelectric materials like PVDF, $BaTiO_3$, PZT, etc. were used initially for NCFET design [38–40,62]. However, apart from having benefits such as being simple to fabricate, these materials are too thick (greater than 10 nm) for outstanding ferroelectric characteristics, making them incompatible with lower technology nodes, and there are also drawbacks in lead use restrictions as well as being incompatible with existing CMOS processes which restrict their uses. Later, hafnium-based doped or undoped ferroelectric materials were developed to address technical difficulties. Lee et al. reported that the $HfZrO_x$ ferroelectric layer with a thickness of 1 nm could be used to validate the steep switching characteristic of NCFETs [63]. Also, for lower technology nodes, transistors' HfO_2 thickness can reach < 1 nm. Further research has revealed that HfO_2 can increase NCFET CMOS compatibility [64,65]. Issues such as traps, leakage, and domain formation [66–69] in hafnium oxide-based ferroelectric layers need to be addressed for proper NCFET device operation. In a nutshell, diverse studies have been done in explaining the negative capacitance phenomenon for steep-switching applications. Controlling hysteresis in NCFET, CMOS-compatible ferroelectric oxides, capacitor matching of ferroelectric and dielectric for stable operations, and proper compact modeling of NCFET are needed before moving to commercialization.

6.4 NCFET BASIC OPERATION

The baseline FET's gate stack is sandwiched with ferroelectric material to form the NCFET structure. The baseline FET could be a MOSFET or a FinFET, dependent on the application. Ferroelectric materials exhibit negative capacitance properties under specific conditions, which improve overall transistor performance. As shown in Figure 6.3, in the literature there are two types of NCFET device structures studied, namely metal ferroelectric metal insulator-semiconductor (M-F-M-I-S) and metal ferroelectric insulator semiconductor (M-F-I-S) [70]. In the case of the M-F-M-I-S structure [71,72], the internal metal layer divides the ferroelectric capacitor and the baseline MOSFET/FinFET into two separate circuit elements connected through the wire and is simulated using a lumped modeling approach by solving the self-consistent Landau-Devonshire (L-D) model of ferroelectric with the MOSFET/FinFET physics. However in the M-F-I-S [73] NCFET structure, the ferroelectric–oxide interface internal voltage varies from source to drain in a longitudinal direction for non-zero drain bias. Therefore, either the full analytical approach, distributed modeling approach, or segmentation approach [74–76] is used for evaluating the channel current. Compared to the M-F-I-S device structure, the M-F-M-I-S structure may offer a higher ON current and make the device less susceptible to hysteresis. Numerous studies [69,77] have suggested that the internal metal gate, however, destabilizes the negative capacitance state because of ferroelectric leakage current and domain formation. The internal metal gate of NCFETs also serves as an additional point of traps, which affects the reliability of NCFETs [77–80].

It can be mathematically deduced that SS can be divided into two parts, namely transport factor (q) and body factor (p), and is shown by Equations 6.8 and 6.9, which assists in explicating the advantages of NC integration in reducing SS. NCFET attains an SS of 60 mV/decade by lowering the body factor in the manner described in Equations 6.10 and 6.11, which further divides the body factor into the inverse of the voltage amplification factor (A_G) multiplied by m. In a conventional FET, factor 'p' is often larger than 1 for positively connected series gate oxide capacitance, and

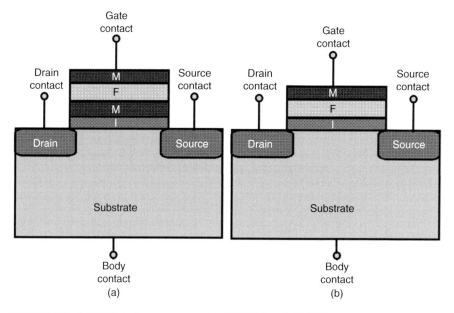

FIGURE 6.3 NCFET device structure: (a) M-F-M-I-S and (b) M-F-I-S.

q is constrained by the thermionic limit of 2.3 KBT/q, or 60 mV/decade. However, in NCFET, the total series capacitance formed by the ferroelectric capacitance (C_{FE}) and total internal FET capacitance (C_{INT}) increase, which causes voltage amplification in the $V_{G'S}$ (Figure 6.4). This happens due to the negative value of C_{FE} which increases A_G to greater than 1, although m is always unity. When compared to conventional FET, the performance of NCFET is improved by this voltage amplification in terms of reducing SS to below the Boltzmann limit and increasing the ON current for the identical applied voltage. This is because the combined term $A_G * m > 1$ results in a reduction of term p to < 1, which in turn lowers the overall SS.

$$SS = \frac{\partial V_{GS}}{\partial \log_{10} I_{DS}} = \underbrace{\frac{\partial V_{GS}}{\partial V_{G'S}} * \frac{\partial V_{G'S}}{\partial \varphi_s}}_{p} * \underbrace{\frac{\partial \varphi_s}{\partial \log_{10} I_{DS}}}_{q} \tag{6.8}$$

$$SS = \frac{1}{A_G.m} * \frac{\partial \varphi_s}{\partial \log_{10} I_{DS}} \tag{6.9}$$

$$m = \frac{\partial \varphi_s}{\partial V_{G'S}} = \left(1 + \frac{C_S}{C_{OX}}\right)^{-1} < 1 \tag{6.10}$$

$$A_G = \frac{\partial V_{G'S}}{\partial V_{GS}} = \left(1 + \frac{C_{INT}}{|C_{FE}|}\right)^{-1} < 1 \text{ for } C_{FE} < 0 \tag{6.11}$$

According to Reference [33], NC instability can be stabilized by connecting a positive dielectric capacitance to the ferroelectric in series, making the capacitance of the entire system positive. A baseline FET's internal gate has positive capacitance (C_{INT}) in the context of an NCFET and is connected in series with the ferroelectric material's capacitance. This makes the total gate capacitance $(1/(C_{INT}^{-1} + C_{FE}^{-1}))$ positive and stabilizes the unstable NC. The condition $|C_{FE}| > C_{INT}$ should follow for capacitor matching. Also, it is evident from Figure 6.4 that $|C_{FE}|$ and C_{INT} are dependent

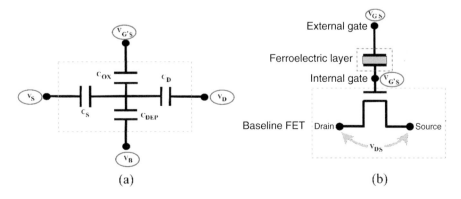

(a)   (b)

FIGURE 6.4   NCFET general circuit equivalent: (a) capacitance model and (b) schematic.

on drain, gate, and body voltages (V_D, V_G, V_B). Violation of the aforementioned condition causes the electrical properties of the device to behave in a hysteresis manner for any combination of V_G and V_D [81]. Because $\left|C_{FE}\right|$ varies in an inverse proportion with T_{FE}, it is necessary to control T_{FE} in order to avoid these conflicts. The remnant polarization (P_r) and coercive field (E_c) of the ferroelectric material also affect $\left|C_{FE}\right|$. The $\left|C_{FE}\right|$ value is increased by ferroelectric materials with high P_r and low E_c values [82]. This is required to maintain a hysteresis-free condition ($|C_{FE}| > C_{INT}$) within the range of applied operating voltages. Also, if C_{INT} and C_{FE} are not matched properly due to selecting low T_{FE}, high P_r, and low E_C, the gain due to ferroelectric decreases, as shown in Equation 6.11. Therefore, in order to achieve higher NC gain while avoiding hysteresis, which is undesirable for logic applications, critical values of E_c, P_r, and T_{FE} must be carefully chosen. Awadhiya et al. [83–87] have shown mathematically that for any general NCFET [Figure 6.4(b)], the effects due to drain (D) and gate (G) in the internal gate (G') are needed. Using the superposition theorem to solve the capacitive model shown in Figure 6.4(a) we get Equations 6.12–6.15 where A_D is defined as a drain coupling factor, A_G is the differential gain or amplification factor, and A_B is the bulk coupling factor for any NCFET device and is given by:

$$V_{G'S} = A_G V_{GS} + A_D V_{DS} + A_B V_{BS} \tag{6.12}$$

$$A_G = \frac{\partial V_{G'S}}{\partial V_{GS}} = \frac{|C_{FE}|}{|C_{FE}| - C_{FET}} \tag{6.13}$$

$$A_D = \frac{\partial V_{G'S}}{\partial V_{DS}} = \frac{C_D}{|C_{FE}| - C_{FET}} \tag{6.14}$$

$$A_B = \frac{\partial V_{G'S}}{\partial V_{BS}} = \frac{C_{DEP}}{|C_{FE}| - C_{FET}} \tag{6.15}$$

These equations describe the general mathematical formulations to show the behavior of any NCFET and have been able to explain its unique features.

6.5   NCFET FEATURES

Apart from reducing SS to below 60 mV/decade, NCFET has some other features which could be a benefit for low-power applications and these have been discussed in many works. Ferroelectric thickness (T_{FE}), E_C, and P_r are critical parameters for any NCFET. Optimization of these parameters for a particular technology is required to obtain hysteresis-free NCFET or hysteretic FEFET which then can be utilized for ultra-low-power logic, analog, and memory applications. To date, it has been discussed in various articles [81,82] that due to the voltage amplification effect of negative capacitance, NCFETs show improved ON current and decreased OFF current as compared to conventional FETs, which makes them a better choice of device for ultra-low-power

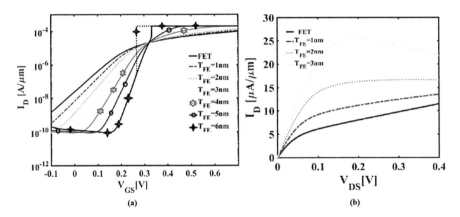

FIGURE 6.5 (a) NCFET general transfer characteristics showing negative DIBL, GIDL, incremented ON current, and decremented OFF current, and (b) output characteristics showing enhanced ON current and NDR with varying T_{FE} in 14 nm FinFET technology.

applications. The general transfer characteristics of the n-type NCFETs with different ferroelectric thicknesses (T_{FE}) are shown in Figure 6.5(a). It can be seen that NCFET has a greater I_{ON}/I_{OFF} ratio compared to the conventional FET. With rising T_{FE}, these benefits become more obvious. Moreover, hysteresis in the output characteristics seems to occur at $T_{FE} = 6$ nm. This is because of an imbalance between the internal total baseline FET capacitance and ferroelectric capacitance (discussed in Section 6.5), which should be avoided if such an NCFET device is to be used for logic applications [60,92]. As shown in Figure 6.5(a), the drain-induced barrier lowering (DIBL) factor in NCFET is negative, hence the threshold voltage rises as the drain bias is enhanced, which has been verified by various literature works [88–90]. The opposite of how FET works where threshold voltage reduces with increasing drain bias, NCFET's barrier height rises as the drain voltage rises, translating into an increase in threshold voltage and resulting in a negative DIBL factor, and this phenomenon is termed DIBR (drain-induced barrier rising) or negative DIBL. Furthermore, gate-induced drain lowering (GIDL) increases with increasing T_{FE} [93,94]. Also, in the output characteristics [Figure 6.5(b)] of NCFET with increasing T_{FE} it can be seen that negative differential resistance (NDR) occurs where the ON current decreases with an increment in the drain to source voltage. This feature also has been verified by various research groups [81,89,91].

6.6 MODELING APPROACHES: NCFET

To evaluate the NCFET's unique features for low-power applications including devices, circuits, systems, etc. lots of compact, analytical modeling approaches and the TCAD model have been proposed and used [71,95–100]. This section discusses two of them, i.e., the Landau-Khalatnikov (L-K)/Landau-Devonshire (L-D) model for ferroelectrics and the Prieasch model of ferroelectrics. In the L-K compact modeling approach, the single domain time-dependent L-K/L-D model of ferroelectric given by Equation 6.16 is coupled with the industry standard FinFET/MOSFET model (BSIM model) [101]. In Equation 6.16 α, β, and γ are materials with specific Landau coefficients and the kinetic coefficient ρ relates with the time constant associated with ferroelectric polarization change.

$$V_{FE} = \left(2\alpha Q_G + 4\beta Q_G^3 + 6\gamma Q_G^5 + \rho \frac{dQ_G}{dt} \right) T_{FE} \qquad (6.16)$$

The gate charge (Q_G) is first retrieved from the calibrated BSIM model in the self-consistent strategy. Then, the Q_G value is applied to the parameter Q_G in the Verilog-A written L-K equation (Equation 6.16). The internal metal voltage given by $V_{G'S} = V_{GS} - V_{FE}$ is the outcome of the L-K model. This modeling strategy relies on a set of assumptions, such as taking the ferroelectric mono-domain formation into account, assuming polarization $P = Q_G - \varepsilon_o * V_{FE}/T_{FE} \approx Q_G$. Initially, the primary issue with the L-K theory of ferroelectric phase transition model is that the S-shaped P–E curve lacks experimental validation. However, later, presented techniques [102] for detecting steady-state negative capacitance and S-curves that were validated by experiment, gave credence to the use of the L-K modeling approach. Contrary to other techniques, this model can capture all of the distinctive characteristics of the NCFET despite its clear and simple methodology [103]. The experimental findings of NCFETs with ferroelectric thicknesses as small as 1.5 nm and 3 nm [63] could also be replicated by the L-K model. Additionally, the NC effect on its own has no additional delay penalty [104]. In contrast to L-K model, the Preisach model [105,106] is based on an integration of the baseline FET model with the multi-domain Preisach theory for ferroelectric. Initially, the L-K model was used, this has some limitations of not capturing hysteresis properly and assuming mono-domain in ferroelectrics, while some groups have recently shown that the Prieasch model of negative capacitance includes more accurate modeling of hysteresis by capturing the multi-domain effect [107,108] in ferroelectric for NCFET, specially FEFET. Both of these modeling approaches are acceptable and have been able to provide experimental evidence of NCFET. The L-K model is generally used for logic applications, while the Prieasch model is used for memory applications.

6.7 CONCLUSION

Considering the pressing need for a new low-power switch, NCFET research after its conceptual proposal by Salahuddin et al. is legitimately deserving of the extensive coverage and in-depth analysis it has received from the device physics community over the past ten years.

This chapter summarizes the basics of NCFET, starting from the initial concept to the recent modifications and its matures current status. The utilization of ferroelectric with the gate stack of conventional FET to achieve sub-60 mV/decade is the primary focus of this fast Landau switch for low-power electronics. Moreover, NCFET achieves this by reducing the threshold voltage, increasing ON-to-OFF current ratio, being hysteresis-free, and achieving supply voltage scaling. However, this comes at the cost of increased total gate capacitance and an added barrier at the drain side due to a decrement in the internal gate voltage at higher drain voltage. At high ferroelectric thickness and low gate-to-source voltage, NDR is seen in the NCFET, which can be useful for analog applications. Internal voltage amplification not only reduces SS but also shows NDIBL with suppressed short channel effect, contrary to the MOSFET. The discovery of ferroelectricity in HfO_2 doped/undoped CMOS-compatible material provides additional interest to the NCFET concept towards commercialization and avoids the problem of ultra scaling in conventional ferroelectric devices for below 5 nm. Optimizations of ferroelectric thickness, coercive field, and polarization are the key parameters to achieve performance benefits in NCFET. Research efforts have been aimed at various device structure configurations, namely MFMIS and MFIS. MFIS NCFET device structure provides additional practical benefits compared with MFMIS. Many compact, TCAD, analytical models have been developed using L-K theory and have been able to explain various experimental phenomena of multidomain switching and hysteresis in NCFET. Quasi-static NCFET is well explained using the L-K model, while the Prieasch model is able to explain FEFET.

6.8 FUTURE SCOPE

Lots of works have been published on NCFET conceptual models but there remain some research gaps which need to be filled before moving to commercialization. The prime focus of NCFET relies on the reduction of SS, which creates a confusing scenario in this field. The frequency response of NCFET needs to be addressed in the future to validate its limit and speed of operations. Studying the reliability of NCFET devices includes threshold voltage and charge accumulation stability, ferroelectric-dielectric heterostructure, and breakdown strength. Also, the aggressive scaling below 5 nm node occurs recently and in the future to provide extremely thin ferroelectric dimensions is itself a challenge while maintaining proper device operation. Proper validation of various physics models and theories for quasi-static NC remains unresolved and debatable. It is too early to use HfO_2-based ferroelectric, with a lot of research needed to ensure its compatibility, reliability, domain dynamics, etc.

REFERENCES

1. Theis, T.N., & Solomon, P.M. (2010) It's time to reinvent the transistor! *Science* 327: 1600–1601.
2. Ionescu, A.M., & Riel, H.E. (2011) Tunnel field-effect transistors as energy-efficient electronic switches. *Nature* 479: 329–337.
3. Shukla, N., Thathachary, A.V., Agrawal, A., Paik, H., Aziz, A., Schlom, D.G., Gupta, S.K., Engel-Herbert, R., & Datta, S. (2015) A steep-slope transistor based on abrupt electronic phase transition. *Nature Communications* 6.
4. Yadav, S., Upadhyay, P., Awadhiya, B., & Kondekar, P.N. (2022) Ferroelectric negative-capacitance-assisted phase-transition field-effect transistor. *IEEE Transactions on Ultrasonics, Ferroelectrics, and Frequency Control* 69: 863–869.

5. Yadav, S., Upadhyay, P., Awadhiya, B., & Kondekar, P.N. (2021) Design and analysis of improved phase-transition FinFET utilizing negative capacitance. *IEEE Transactions on Electron Devices* 68: 853–859.

6. Yadav, S., Kondekar, P.N., Upadhyay, P., & Awadhiya, B. (2022) Negative capacitance based phase-transition FET for low power applications: Device-circuit co-design. *Microelectronics Journal* 123: 105411.

7. Valasek, J. (1921) Piezo-electric and allied phenomena in Rochelle salt. *Physical Review* 17: 475–481.

8. Hippel, A.R., Breckenridge, R.G., Chesley, F.G., & Tisza, L. (1946) High dielectric constant ceramics. *Industrial & Engineering Chemistry* 38: 1097–1109.

9. Scott, J.F. (2007) Applications of modern ferroelectrics. *Science* 315: 954–959.

10. Schenk, T., Mueller, S., Schroeder, U., Materlik, R., Kersch, A., Popovici, M.I., Adelmann, C., Elshocht, S.V., & Mikolajick, T. (2013) Strontium doped hafnium oxide thin films: Wide process window for ferroelectric memories. In: *2013 Proceedings of the European Solid-State Device Research Conference (ESSDERC)*, pp. 260–263.

11. Mueller, S., Mueller, J., Singh, A., Riedel, S., Sundqvist, J., Schroeder, U., & Mikolajick, T. (2012) Incipient ferroelectricity in Al-doped HfO_2 thin films. *Advanced Functional Materials* 22.

12. Polakowski, P., Riedel, S., Weinreich, W., Rudolf, M., Sundqvist, J., Seidel, K., & Muller, J. (2014) Ferroelectric deep trench capacitors based on Al:HfO_2 for 3D non-volatile memory applications. In: *2014 IEEE 6th International Memory Workshop (IMW)*, pp. 1–4.

13. Muller, J., Boscke, T.S., Muller, S., Yurchuk, E., Polakowski, P., Paul, J., Martin, D., Schenk, T., Khullar, K., Kersch, A., Weinreich, W., Riedel, S., Seidel, K., Kumar, A., Arruda, T.M., Kalinin, S.V., Schlosser, T., Boschke, R., van Bentum, R., Schroder, U., & Mikolajick, T. (2013) Ferroelectric hafnium oxide: A CMOS-compatible and highly scalable approach to future ferroelectric memories. In: *2013 IEEE International Electron Devices Meeting*, pp. 10.8.1–10.8.4.

14. Müller, J., Schröder, U., Böscke, T.S., Müller, I., Böttger, U., Wilde, L., Sundqvist, J., Lemberger, M., Kücher, P., Mikolajick, T., & Frey, L. (2011) Ferroelectricity in yttrium-doped hafnium oxide. *Journal of Applied Physics* 110: 114113.

15. Olsen, T., Schröder, U., Müller, S., Krause, A., Martin, D., Singh, A., Müller, J., Geidel, M., & Mikolajick, T. (2012) Co-sputtering yttrium into hafnium oxide thin films to produce ferroelectric properties. *Applied Physics Letters* 101: 082905.

16. Mueller, S., Adelmann, C., Singh, A., Elshocht, S.V., Schroeder, U., & Mikolajick, T. (2012) Ferroelectricity in Gd-doped HfO_2 thin films. *ECS Journal of Solid State Science and Technology* 1.

17. Müller, J., Böscke, T.S., Bräuhaus, D., Schröder, U., Böttger, U., Sundqvist, J., Kücher, P., Mikolajick, T., & Frey, L. (2011) Ferroelectric Zr0.5Hf0.5O2 thin films for non-volatile memory applications. *Applied Physics Letters* 99: 112901.

18. Lomenzo, P.D., Zhao, P., Takmeel, Q., Moghaddam, S., Nishida, T., Nelson, M.A., Fancher, C.M., Grimley, E.D., Sang, X., Lebeau, J.M., & Jones, J.L. (2014) Ferroelectric phenomena in Si-doped HfO_2 thin films with TiN and Ir electrodes. *Journal of Vacuum Science & Technology. B. Nanotechnology and Microelectronics: Materials, Processing, Measurement, and Phenomena* 32.

19. Park, M.H., Kim, H., Kim, Y.J., Lee, W., Kim, H.K., & Hwang, C.S. (2013) Effect of forming gas annealing on the ferroelectric properties of Hf0.5Zr0.5O2 thin films with and without Pt electrodes. *Applied Physics Letters* 102: 112914.

20. Park, M.H., Kim, H., Kim, Y.J., Lee, W., Moon, T., & Hwang, C.S. (2013) Evolution of phases and ferroelectric properties of thin Hf0.5Zr0.5O2 films according to the thickness and annealing temperature. *Applied Physics Letters* 102: 242905.
21. Schroeder, U., Yurchuk, E., Müller, J., Martin, D., Schenk, T., Polakowski, P., Adelmann, C., Popovici, M.I., Kalinin, S.V., & Mikolajick, T. (2014) Impact of different dopants on the switching properties of ferroelectric hafniumoxide. *Japanese Journal of Applied Physics* 53.
22. Böscke, T.S., Teichert, S., Bräuhaus, D., Müller, J., Schröder, U., Böttger, U., & Mikolajick, T. (2011) Phase transitions in ferroelectric silicon doped hafnium oxide. *Applied Physics Letters* 99: 112904.
23. Zhou, D., Müller, J., Xu, J., Knebel, S., Bräuhaus, D., & Schröder, U. (2012) Insights into electrical characteristics of silicon doped hafnium oxide ferroelectric thin films. *Applied Physics Letters* 100: 082905.
24. Zhou, D., Xu, J., Li, Q., Guan, Y., Cao, F., Dong, X., Müller, J., Schenk, T., & Schröder, U. (2013) Wake-up effects in Si-doped hafnium oxide ferroelectric thin films. *Applied Physics Letters* 103: 192904.
25. Yurchuk, E., Müller, J., Knebel, S., Sundqvist, J., Graham, A.P., Melde, T., Schröder, U., & Mikolajick, T. (2013) Impact of layer thickness on the ferroelectric behaviour of silicon doped hafnium oxide thin films. *Thin Solid Films* 533: 88–92.
26. Polakowski, P., & Müller, J. (2015). Ferroelectricity in undoped hafnium oxide. *Applied Physics Letters* 106: 232905.
27. Kim, K., Park, M.H., Kim, H., Kim, Y.J., Moon, T., Lee, Y., Hyun, S., Gwon, T., & Hwang, C.S. (2016) Ferroelectricity in undoped-HfO$_2$ thin films induced by deposition temperature control during atomic layer deposition. *Journal of Materials Chemistry C* 4: 6864–6872.
28. Khan, A.I., Chatterjee, K., Wang, B., Drapcho, S., You, L., Serrao, C.R., Bakaul, S.R., Ramesh, R., & Salahuddin, S.S. (2014) Negative capacitance in a ferroelectric capacitor. *Nature Materials* 14(2): 182–186.
29. Hoffmann, M., Fengler, F.P., Herzig, M., Mittmann, T., Max, B., Schroeder, U., Negrea, R.F., Lucian, P., Slesazeck, S., & Mikolajick, T. (2019) Unveiling the double-well energy landscape in a ferroelectric layer. *Nature* 565: 464–467.
30. Landauer, R. (1976). Can capacitance be negative. *Collective Phenomena* 2: 167.
31. Khan, A.I., Bhowmik, D., Yu, P., Kim, S.J., Pan, X., Ramesh, R., & Salahuddin, S.S. (2011) Experimental evidence of ferroelectric negative capacitance in nanoscale heterostructures. *Applied Physics Letters* 99: 113501.
32. Thomas, S. (2019). Negative capacitance found. *Nature Electronics* 2: 51.
33. Salahuddin, S.S., & Datta, S. (2008) Use of negative capacitance to provide voltage amplification for low power nanoscale devices. *Nano Letters* 8(2): 405–410.
34. Zhirnov, V.V., & Cavin, R.K. (2008) Nanoelectronics: negative capacitance to the rescue? *Nature Nanotechnology* 3(2): 77–78.
35. Catalán, G., Jiménez, D., & Gruverman, A. (2015) Ferroelectrics: Negative capacitance detected. *Nature Materials* 14(2): 137–139.
36. Appleby, D.J., Ponon, N., Kwa, K.S., Zou, B., Petrov, P.K., Wang, T., Alford, N.M., & O'Neill, A. (2014) Experimental observation of negative capacitance in ferroelectrics at room temperature. *Nano Letters* 14(7): 3864–3868.
37. Zhao, Z., Buscaglia, V., Viviani, M., Buscaglia, M.T., Mitoseriu, L., Testino, A., Nygren, M., Johnsson, M., & Nanni, P. (2004) Grain-size effects on the ferroelectric behavior of dense nanocrystalline BaTiO$_3$ ceramics. *Physical Review B* 70: 024107.

38. Gao, W., Khan, A.I., Marti, X., Nelson, C., Serrao, C.R., Ravichandran, J., Ramesh, R., & Salahuddin, S.S. (2014) Room-temperature negative capacitance in a ferroelectric-dielectric superlattice heterostructure. *Nano Letters* 14(10): 5814–5819.

39. Ku, H., & Shin, C. (2017) Transient response of negative capacitance in P(VDF0.75-TrFE0.25) organic ferroelectric capacitor. *IEEE Journal of the Electron Devices Society* 5: 232–236.

40. Jo, J., Choi, W.Y., Park, J., Shim, J.W., Yu, H., & Shin, C. (2015). Negative capacitance in organic/ferroelectric capacitor to implement steep switching MOS devices. *Nano Letters* 15(7): 4553–4556.

41. Jo, J., & Shin, C. (2016) Negative capacitance field effect transistor with hysteresis-free sub-60-mV/decade switching. *IEEE Electron Device Letters* 37: 245–248.

42. Dasgupta, S., Rajashekhar, A., Majumdar, K., Agrawai, N., Razavieh, A., Troiier-McKinstry, S., & Datta, S. (2015) Sub-kT/q switching in strong inversion in PbZr0.52Ti0.48O3 gated negative capacitance FETs. *IEEE Journal on Exploratory Solid-State Computational Devices and Circuits* 1: 43–48.

43. Sharma, P., Tapily, K., Saha, A.K., Zhang, J., Shaughnessy, A., Aziz, A., Snider, G.L., Gupta, S.K., Clark, R.D., & Datta, S. (2017) Impact of total and partial dipole switching on the switching slope of gate-last negative capacitance FETs with ferroelectric hafnium zirconium oxide gate stack. *2017 Symposium on VLSI Technology* T154–T155.

44. Kobayashi, M., & Hiramoto, T. (2016) On device design for steep-slope negative-capacitance field-effect-transistor operating at sub-0.2V supply voltage with ferroelectric HfO$_2$ thin film. *AIP Advances* 6: 025113.

45. Awadhiya, B., Kondekar, P.N., & Meshram, A.D. (2018) Analogous behavior of FE-DE heterostructure at room temperature and ferroelectric capacitor at Curie temperature. *Superlattices and Microstructures* 123: 306–310.

46. Kondekar, P.N., & Awadhiya, B. (2017). Effect of parameter variation in UTBB FDSOINCFET. In: *2017 Joint IEEE International Symposium on the Applications of Ferroelectric (ISAF)/International Workshop on Acoustic Transduction Materials and Devices (IWATMD)/Piezoresponse Force Microscopy (PFM)*, pp. 45–47.

47. Krivokapic, Z., Rana, U., Galatage, R., Razavieh, A., Aziz, A., Liu, J.K., Shi, J., Kim, H., Sporer, R., Serrao, C.R., Busquet, A., Polakowski, P., Müller, J., Kleemeier, W., Jacob, A., Brown, D.E., Knorr, A., Carter, R., & Banna, S. (2017) 14nm Ferroelectric FinFET technology with steep subthreshold slope for ultra low power applications. in: *2017 IEEE International Electron Devices Meeting (IEDM)*, pp. 15.1.1–15.1.4.

48. Hsu, C., Pan, C., & Naeemi, A. (2018) Performance analysis and enhancement of negative capacitance logic devices based on internally resistive ferroelectrics. *IEEE Electron Device Letters* 39: 765–768.

49. Li, K., Chen, P., Lai, T., Lin, C., Cheng, C., Chen, C., Wei, Y., Hou, Y.F., Liao, M.H., Lee, M., Chen, M., Sheih, J., Yeh, W.K., Yang, F., Salahuddin, S.S., & Hu, C.C. (2015) Sub-60mV-swing negative-capacitance FinFET without hysteresis. In: *2015 IEEE International Electron Devices Meeting (IEDM)*, pp. 22.6.1–22.6.4.

50. Ko, E., Lee, J.W., & Shin, C. (2017) Negative capacitance FinFET with sub-20-mV/decade subthreshold slope and minimal hysteresis of 0.48 V. *IEEE Electron Device Letters* 38: 418–421.

51. Mcguire, F., Cheng, Z., Price, K., & Franklin, A.D. (2016) Sub-60 mV/decade switching in 2D negative capacitance field-effect transistors with integrated ferroelectric polymer. *Applied Physics Letters* 109: 093101.

52. Mcguire, F., Lin, Y., Rayner, B., & Franklin, A.D. (2017) MoS$_2$ negative capacitance FETs with CMOS-compatible hafnium zirconium oxide. In: *2017 75th Annual Device Research Conference (DRC)*, pp. 1–2.

53. Mcguire, F., Lin, Y., Price, K., Rayner, G.B., Khandelwal, S., Salahuddin, S.S., & Franklin, A.D. (2017) Sustained sub-60 mV/decade switching via the negative capacitance effect in MoS_2 transistors. *Nano Letters* 17(8): 4801–4806.

54. Ionescu, A.M. (2017) Negative capacitance gives a positive boost. *Nature Nanotechnology* 13: 7–8.

55. Masuduzzaman, M., & Alam, M.A. (2014). Effective nanometer airgap of NEMS devices using negative capacitance of ferroelectric materials. *Nano Letters* 14(6): 3160–3165.

56. Choe, K., & Shin, C. (2017) Adjusting the operating voltage of a nanoelectromechanical relay using negative capacitance. *IEEE Transactions on Electron Devices* 64: 5270–5273.

57. Srimani, T., Hills, G., Bishop, M.D., Radhakrishna, U., Zubair, A., Park, R.S., Stein, Y., Palacios, T., Antoniadis, D.A., & Shulaker, M.M. (2018) Negative capacitance carbon nanotube FETs. *IEEE Electron Device Letters* 39: 304–307.

58. Ota, H., Migita, S., Hattori, J., Fukuda, K., & Toriumi, A. (2016) Design and simulation of steep-slope silicon-on-insulator FETs using negative capacitance: Impact of buried oxide thickness and remnant polarization. In: *2016 IEEE 16th International Conference on Nanotechnology (IEEE-NANO)*, pp. 770–772.

59. Naber, R.C., Tănase, C., Blom, P.W., Gelinck, G.H., Marsman, A.W., Touwslager, F.J., Setayesh, S., & de Leeuw, D.M. (2005) High-performance solution-processed polymer ferroelectric field-effect transistors. *Nature Materials* 4: 243–248.

60. Jaisawal, R.K., Kondekar, P.N., Yadav, S., Upadhyay, P., Awadhiya, B., & Rathore, S. (2021) Insights into the operation of negative capacitance FinFET for low power logic applications. *Microelectronics Journal* 119: 105321.

61. Saeidi, A., Jazaeri, F., Bellando, F., Stolichnov, I., Enz, C.C., & Ionescu, A.M. (2017) Negative capacitance field effect transistors; capacitance matching and non-hysteretic operation. In: *2017 47th European Solid-State Device Research Conference (ESSDERC)*, pp. 78–81.

62. Sharma, P., Zhang, J., Ni, K., & Datta, S. (2018) Time-resolved measurement of negative capacitance. *IEEE Electron Device Letters* 39: 272–275.

63. Lee, M.H., Fan, S., Tang, C., Chen, P., Chou, Y., Chen, H., Kuo, J.Y., Xie, M.J., Liu, S., Liao, M.H., Jong, C.A., Li, K., Chen, M., & Liu, C.W. (2016) Physical thickness 1.x nm ferroelectric HfZrOx negative capacitance FETs. In: *2016 IEEE International Electron Devices Meeting (IEDM)*, pp. 12.1.1–12.1.4.

64. Sharma, A., & Roy, K. (2017) Design space exploration of hysteresis-free HfZrOx-based negative capacitance FETs. *IEEE Electron Device Letters* 38: 1165–1167.

65. Ko, E., Lee, H., Goh, Y., Jeon, S., & Shin, C. (2017) Sub-60-mV/decade negative capacitance FinFET with sub-10-nm hafnium-based ferroelectric capacitor. *IEEE Journal of the Electron Devices Society* 5: 306–309.

66. Kasamatsu, S., Watanabe, S., Hwang, C.S., & Han, S. (2016) Emergence of negative capacitance in multidomain ferroelectric–paraelectric nanocapacitors at finite bias. *Advanced Materials* 28.

67. Shin, Y., Grinberg, I., Chen, I., & Rappe, A.M. (2007) Nucleation and growth mechanism of ferroelectric domain-wall motion. *Nature* 449: 881–884.

68. Li, Y., Hu, S., Liu, Z., & Chen, L. (2002) Effect of substrate constraint on the stability and evolution of ferroelectric domain structures in thin films. *Acta Materialia* 50: 395–411.

69. Khan, A.I., Radhakrishna, U., Chatterjee, K., Salahuddin, S.S., & Antoniadis, D.A. (2016) Negative capacitance behavior in a leaky ferroelectric. *IEEE Transactions on Electron Devices* 63: 4416–4422.

70. Lee, S., Chen, H., Shen, C., Kuo, P., Chung, C., Huang, Y., Chen, H., & Chao, T. (2019) Experimental demonstration of performance enhancement of MFMIS and MFIS for 5-nm × 12.5-nm poly-Si nanowire gate-all-around negative capacitance FETs featuring seed-layer and PMA-free process. In: *2019 Silicon Nanoelectronics Workshop (SNW)*, pp. 1–2.

71. Aziz, A., Ghosh, S., Datta, S., & Gupta, S.K. (2016) Physics-based circuit-compatible SPICE model for ferroelectric transistors. *IEEE Electron Device Letters* 37: 805–808.

72. Li, Y., Lian, Y., Yao, K., & Samudra, G.S. (2015) Evaluation and optimization of short channel ferroelectric MOSFET for low power circuit application with BSIM4 and Landau theory. *Solid-state Electronics* 114: 17–22.

73. Pahwa, G., Dutta, T., Agarwal, A., & Chauhan, Y.S. (2017) Compact model for ferroelectric negative capacitance transistor with MFIS structure. *IEEE Transactions on Electron Devices* 64: 1366–1374.

74. Pahwa, G., Dutta, T., Agarwal, A., & Chauhan, Y.S. (2018) Physical insights on negative capacitance transistors in nonhysteresis and hysteresis regimes: MFMIS versus MFIS structures. *IEEE Transactions on Electron Devices* 65: 867–873.

75. Duarte, J.P., Khandelwal, S., Khan, A.I., Sachid, A.B., Lin, Y., Chang, H., Salahuddin, S.S., & Hu, C.C. (2016) Compact models of negative-capacitance FinFETs: Lumped and distributed charge models. In: *2016 IEEE International Electron Devices Meeting (IEDM)*, pp. 30.5.1–30.5.4.

76. Jiménez, D., Miranda, E.A., & Godoy, A. (2010) Analytic model for the surface potential and drain current in negative capacitance field-effect transistors. *IEEE Transactions on Electron Devices* 57: 2405–2409.

77. Saha, A.K., & Gupta, S.K. (2019) Multi-domain negative capacitance effects in metal-ferroelectric-insulator-semiconductor/metal stacks: A phase-field simulation based study. *Scientific Reports* 10.

78. Tang, Y., Su, C.J., Wang, Y., Kao, K., Wu, T., Sung, P., Hou, F., Wang, C., Yeh, M.S., Lee, Y., Wu, W., Huang, G., Shieh, J.M., Yeh, W.K., & Wang, Y. (2018) A comprehensive study of polymorphic phase distribution of ferroelectric-dielectrics and interfacial layer effects on negative capacitance FETs for sub-5 nm node. In: *2018 IEEE Symposium on VLSI Technology*, pp. 45–46.

79. Fan, C., Tu, C., Lin, M., Chang, C., Cheng, C., Chen, Y., Liou, G., Liu, C., Chou, W., & Hsu, H. (2018) Interface engineering of ferroelectric negative capacitance FET for hysteresis-free switch and reliability improvement. In: *2018 IEEE International Reliability Physics Symposium (IRPS)*, pp. P-TX.8-1–P-TX.8-5.

80. Liu, C., Chen, H., Hsu, C., Fan, C., Hsu, H., & Cheng, C. (2019) Negative capacitance CMOS field-effect transistors with non-hysteretic steep sub-60mV/dec swing and defect-passivated multidomain switching. In: *2019 Symposium on VLSI Technology*, pp. T224–T225.

81. Pahwa, G., Dutta, T., Agarwal, A., Khandelwal, S., Salahuddin, S.S., Hu, C.C., & Chauhan, Y.S. (2016) Analysis and compact modeling of negative capacitance transistor with high ON-current and negative output differential resistance—Part I: Model description. *IEEE Transactions on Electron Devices* 63: 4981–4985.

82. Pahwa, G., Dutta, T., Agarwal, A., & Chauhan, Y.S. (2016) Designing energy efficient and hysteresis free negative capacitance FinFET with negative DIBL and 3.5X ION using compact modeling approach. In: *ESSCIRC Conference 2016: 42nd European Solid-State Circuits Conference*, pp. 49–54.

83. Awadhiya, B., Kondekar, P.N., & Meshram, A.D. (2019) Investigating undoped HfO_2 as ferroelectric oxide in leaky and non-leaky FE–DE heterostructure. *Transactions on Electrical and Electronic Materials* 1–6.

84. Awadhiya, B., Kondekar, P.N., & Meshram, A.D. (2018) Passive voltage amplification in non-leaky ferroelectric–dielectric heterostructure. *Micro & Nano Letters* 13(10): 1399–1403.

85. Awadhiya, B., Kondekar, P.N., Yadav, S., Upadhyay, P., Jaisawal, R.K., & Rathore, S. (2021) Effect of scaling on passive voltage amplification in FE-DE hetero structure. In: *2021 International Conference on Control, Automation, Power and Signal Processing (CAPS)*, pp. 1–4.

86. Awadhiya, B., Kondekar, P.N., & Meshram, A.D. (2019) Effect of ferroelectric thickness variation in undoped HfO_2-based negative-capacitance field-effect transistor. *Journal of Electronic Materials* 48: 6762–6770.

87. Awadhiya, B., Yadav, S., Upadhyay, P., & Kondekar, P.N. (2022) Effect of back gate biasing in negative capacitance field effect transistor. *Micro and Nanostructures* 166: 207226.

88. Awadhiya, B., Kondekar, P.N., Yadav, S., & Upadhyay, P. (2020) Insight into threshold voltage and drain induced barrier lowering in negative capacitance field effect transistor. *Transactions on Electrical and Electronic Materials* 22: 267–273.

89. Liang, Y., Li, X., Gupta, S.K., Datta, S., & Narayanan, V. (2018) Analysis of DIBL effect and negative resistance performance for NCFET based on a compact SPICE model. *IEEE Transactions on Electron Devices* 65: 5525–5529.

90. Seo, J., Lee, J., & Shin, M. (2017) Analysis of drain-induced barrier rising in short-channel negative-capacitance FETs and its applications. *IEEE Transactions on Electron Devices* 64: 1793–1798.

91. Awadhiya, B., Kondekar, P.N., & Meshram, A.D. (2019) Understanding negative differential resistance and region of operation in undoped HfO_2-based negative capacitance field effect transistor. *Applied Physics A* 125: 1–7.

92. Liang, Y., Zhu, Z., Li, X., Gupta, S.K., Datta, S., & Narayanan, V. (2020) Mismatch of ferroelectric film on negative capacitance FETs performance. *IEEE Transactions on Electron Devices* 67: 1297–1304.

93. Gaidhane, A.D., Pahwa, G., Verma, A., & Chauhan, Y.S. (2020) Gate-induced drain leakage in negative capacitance FinFETs. *IEEE Transactions on Electron Devices* 67: 802–809.

94. Min, J., Choe, G., & Shin, C. (2020) Gate-induced drain leakage (GIDL) in MFMIS and MFIS negative capacitance FinFETs. *Current Applied Physics* 20: 1222–1225.

95. Pahariya, A., & Dutta, A.K. (2022) A new surface potential-based analytical model for MFIS NCFETs. *IEEE Transactions on Electron Devices* 69: 870–877.

96. Zhao, Y., Li, L., Peng, Y., Li, Q., Yang, G., Chuai, X., Li, Q., Han, G., & Liu, M. (2019) Surface potential-based compact model for negative capacitance FETs compatible for logic circuit: with time dependence and multidomain interaction. In: *2019 IEEE International Electron Devices Meeting (IEDM)*, pp. 7.5.1–7.5.4.

97. Lee, H., Yoon, Y., & Shin, C. (2017) Current-voltage model for negative capacitance field-effect transistors. *IEEE Electron Device Letters* 38: 669–672.

98. Pandey, N., & Chauhan, Y.S. (2020) Analytical modeling of short-channel effects in MFIS negative-capacitance FET including quantum confinement effects. *IEEE Transactions on Electron Devices* 67: 4757–4764.

99. Gaidhane, A.D., Pahwa, G., Verma, A., & Chauhan, Y.S. (2018) Compact modeling of drain current, charges, and capacitances in long-channel gate-all-around negative capacitance MFIS transistor. *IEEE Transactions on Electron Devices* 65: 2024–2032.

100. Huang, X., Chen, X., Li, L., Zhong, H., Jiao, Y., Lin, X., Huang, Q., Zhang, L., & Huang, R. (2021) A dynamic current model for MFIS negative capacitance transistors. *IEEE Transactions on Electron Devices* 68: 3665–3671.

101. Dasgupta, A., & Hu, C.C. (2020) BSIM-CMG compact model for IC CAD: from FinFET to gate-all-around FET technology. *Journal of Microelectronic Manufacturing* 3: 1–10.

102. Yadav, A.K., Nguyen, K.X., Hong, Z., García-Fernández, P., Aguado-Puente, P., Nelson, C.T., Das, S., Prasad, B., Kwon, D.W., Cheema, S.S., Khan, A.I., Hu, C., Íñiguez, J., Junquera, J., Chen, L., Muller, D.A., Ramesh, R., & Salahuddin, S.S. (2019) Spatially resolved steady-state negative capacitance. *Nature* 565: 468–471.

103. Alam, M.A., Si, M., & Ye, P.D. (2019) A critical review of recent progress on negative capacitance field-effect transistors. *Applied Physics Letters* 114: 090401-1–090401-6.

104. Kwon, D.W., Liao, Y., Lin, Y., Duarte, J.P., Chatterjee, K., Tan, A.J., Yadav, A.K., Hu, C.C., Krivokapic, Z., & Salahuddin, S.S. (2018) Response Speed of Negative Capacitance FinFETs. In: *2018 IEEE Symposium on VLSI Technology*, pp. 49–50.

105. Kim, T., del Alamo, J.A., & Antoniadis, D.A. (2020) Dynamics of HfZrO$_2$ ferroelectric structures: experiments and models. In: *2020 IEEE International Electron Devices Meeting (IEDM)*, pp. 21.4.1–21.4.4.

106. Obradovic, B., Rakshit, T., Hatcher, R.M., Kittl, J.A., & Rodder, S.M. (2018) Modeling of negative capacitance of ferroelectric capacitors as a non-quasi static effect. *arXiv: Mesoscale and Nanoscale Physics* 1–8.

107. Singh, K.J., Bulusu, A., & Dasgupta, S. (2020) Multidomain negative capacitance effect in P(VDF-TrFE) ferroelectric capacitor and passive voltage amplification. *IEEE Transactions on Electron Devices* 67: 4696–4700.

108. Ota, H., Ikegami, T., Fukuda, K., Hattori, J., Asai, H., Endo, K., Migita, S., & Toriumi, A. (2018) Multidomain dynamics of ferroelectric polarization and its coherency-breaking in negative capacitance field-effect transistors. In: *2018 IEEE International Electron Devices Meeting (IEDM)*, pp. 9.1.1–9.1.4.

7 Spin-based Magnetic Devices with Spintronics

Asif Rasool[a], Shahnaz Kossar[b], R. Amiruddin[c]
[a]Department of Applied Science, Maulana Mukhtar Ahmad Nadvi Technical Campus, Mansoora, Malegaon-423203
[b]Department of Physics, GNA University, Sri Hargobindgarh, Phagwara-Hoshiarpur Road, Phagwara, Punjab 144401
[c]Department of Physics, B.S. Abdur Rahman Crescent Institute of Science and Technology, Chennai-600048

7.1 INTRODUCTION

With advances in fabrication technology, complementary metal oxide semiconductor (CMOS) technology has undergone dramatic downscaling over the past five decades, as a result of which the performance of integrated circuits (ICs) has steadily improved following Moore's law [1,2]. As a consequence of this law, CMOS transistors became smaller, but their power consumption increased due to the scaling of physical features, while the amount of leakage currents due to quantum tunneling also increased [3]. Furthermore, the memory devices based on CMOS technology require a high amount of power in order to maintain their non-volatility property [4–6]. As a result, researchers are seeking an alternative to CMOS technology, leading to the development of new technologies [7–9].

Spintronics, also referred to as magnetoelectronics, is considered to be one of the most promising future technologies. Spintronic devices make use of the spinning of an electron along with its fundamental electronic charge [10,11]. Elementary particles such as electrons, neutrons, protons, neutrinos, and neutrinos possess an inherent spin property [12]. The two important characteristics of an electron, that is spin orientation and its magnetic moment, are used as state variables, which have the potential to solve many of the problems of charge-based electronic devices [13]. Spintronic technology retaining spin and magnetization of an electron are very unlike conventional CMOS technology where stored charge is lost due to leakage current [14,15].

Static random-access memory (SRAM) and dynamic random-access memory (DRAM) are used for data storage. Table 7.1 shows a comparison of the performance parameters of various memory storage devices. However, SRAM and DRAM memory devices suffer certain limitations which include high power dissipation and high leakage current, depreciation of stored charge, and also require regular refreshing circuits [16,17]. These problems can be overcome with spintronic-based magnetic random access memory (MRAM), such as spin transfer torque (STT) MRAM and spin–orbit torque (SOT) MRAM, due to their high endurance,

DOI: 10.1201/9781003373391-7

103

TABLE 7.1
The comparison of performance parameters of various memory storage devices

Performance parameters	SRAM	DRAM	Flash (NAND	FeRAM	ReRAM	PCRAM	STT-MRAM	SOT-MRAM
Cell size	50–120	6–10	5	15–34	6–10	4–19	6–20	6–20
Read time (ns)	≤ 2	~30	~10^3	~5	1–20	~2	1–20	≤ 10
Write time (ns)	≤ 2	~50	~10^6	~10	~50	~10^2	~10	≤ 10
Power consumption	Low	Low	High	Low	Medium	Low	Low	Low
Endurance (cycles)	10^{16}	10^{16}	10^5	10^{12}	10^6	10^{10}	10^{15}	10^{15}

non-volatile nature, and fast read/write speeds. In the fast-growing fields of big data, artificial intelligence, and information and communication technology (ICT), spintronic devices can have a huge impact [18–20]. This chapter focuses on the spintronic-based magnetic devices and their applications in memory technology.

7.2 SPINTRONIC DEVICES

Figure 7.1 shows the key historical developments in spintronics-based research. The concept of electron spin was proposed in 1925 [21], many years before the first integrated circuit was introduced in 1958 [22,23].

However, because of technological constraints and a lack of understanding, it was not widely used until the discovery of giant magnetoresistance (GMR) in 1988 [24,25]. GMR is regarded as one of the physical breakthroughs that led to the development of spintronics.

The giant magnetoresistance (MGR) effect is the change in electrical resistance of magnetic material upon the application of an external magnetic field [26]. Another milestone in the field of spintronics was the proposal for a spin transistor by Datta and Das in 1990 [27], also known as an electro-optic modulator. However, the spin transistor sparked the development of a wide range of spintronic-based future technologies and concepts. Figure 7.2 illustrates the major classifications of spintronic devices. Among them, spin valve, MTJ, domain wall nanowire, all spin logic devices, and skyrmions are the most promising ones due to their high performance and future scope [26].

7.3 SPIN VALVE

In general, a spin valve is a sample made up of a GMR trilayer which includes two ferromagnetic layers separated by a non-ferromagnetic layer. Of the two ferromagnetic layers, one is magnetically soft and is very sensitive to small applied magnetic fields, whereas the other layer is magnetically hard [28]. One of the ferromagnetic layers has a fixed magnetic orientation called a pinned layer, while the other layer can either have a parallel or antiparallel magnetic orientation called a free layer (FL).

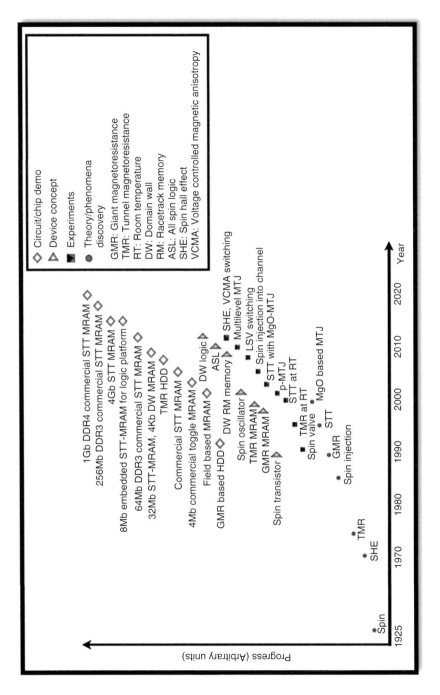

FIGURE 7.1 The key historical developments in spintronics-based research.

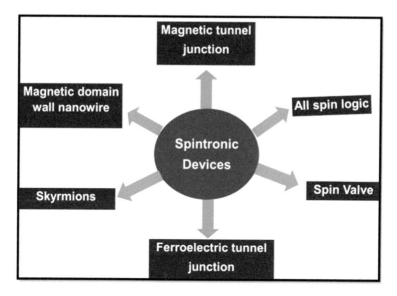

FIGURE 7.2   Major classifications of spintronic devices.

During the giant magnetoresistance effect, the ferromagnetic layers react with the applied magnetic field to change their magnetization relative to each other, as a result the electric conductivity changes. The spin valve is based upon the fact that the spin state of carriers (electrons) align either 'up' or 'down' in ferromagnetic materials upon the application of an external magnetic field [29,30]. A spin valve can be understood by considering the scattering of spin-polarized electrons [31].

Figure 7.3 shows the conduction principle in the spin valve. In ferromagnetic material, the densities of states available for spin-up and spin-down electrons are unequal, while in the case of non-ferromagnetic material the densities of states for spin-up and spin-down electrons are equal. The electrons whose spin is aligned in the same direction as magnetic orientation of the ferromagnetic material layer are called the majority electrons, whereas those electrons whose spin is aligned in the opposite direction as magnetic orientation of the FM layer are called minority electrons [32]. The spin valve has relatively low electrical resistance in a parallel configuration, where majority electrons (spin-up) flow freely through NFM, compared to minority electrons (spin-down). On the other hand, when the spin valve is in antiparallel configuration, the arrangement of density of states causes scattering of both spin-up as well as spin-down electrons, resulting in relatively high resistance of the device [33].

The equation for GMR is given as:

$$\text{GMR} = \frac{\Delta R}{R_p} \qquad (7.1)$$

where ΔR represents the difference of electrical resistance between parallel and anti-parallel configurations of spin arrangements, and R_p denotes the resistance value of

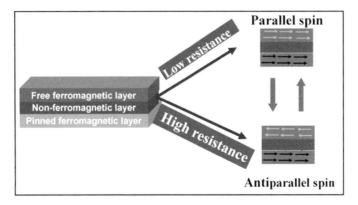

FIGURE 7.3 The conduction principle in the spin valve.

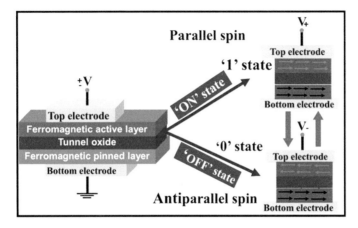

FIGURE 7.4 Schematic illustration of the MRAM device.

the GMR device in parallel configuration. The GMR effect has been utilized in several applications such as GMR sensors, biological applications, and space applications [34].

7.4 MAGNETIC TUNNEL JUNCTION

Figure 7.4 shows the basic structure of a magnetic tunnel junction (MTJ). It consists of two ferromagnetic active layers called the pinned (fixed) and free (storage) layers separated by an intermediate thin insulating oxide or tunnel layer [35].

Depending upon the magnetic orientation of ferromagnetic layers, MTJ is of two types: (i) in-plane MTJ (i-MTJ) and (ii) perpendicular-plane MTJ (p-MTJ). i-MTJ occurs when the magnetic orientation of the FM layers (pinned and fixed layers) is in the plane of the MTJ. On the other hand, if the magnetic orientation of the FM layer is perpendicular to the MTJ plane, then it is p-MTJ [36–38]. Figure 7.5 illustrates the low resistance state (R_p) and high resistance state (R_{AP}) of (a) i-MTJ and (b) p-MTJ.

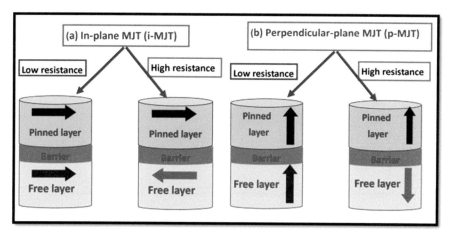

FIGURE 7.5   The low resistance state (R_p) and high resistance state (R_{AP}) of: (a) in-plane MTJ (i-MTJ) and (b) perpendicular-plane MTJ (p-MTJ).

The magnetization of the pinned layer (PL) is fixed and varied for the free (FL) or storage layer [39]. Upon applying an external electric field to the MRAM device, the magnitude of the current passing through the tunneling layer depends on the relative direction (parallel or antiparallel) of magnetization [40]. In the parallel state (P), the resistance offered by the device for the flow of current is reduced, because the magnetic orientation of the free layer (FL) and pinned layer (PL) is in same direction. If, on the other hand, the magnetic orientation of FL is in the opposite direction to the magnetic orientation of the PL, then the device offers a higher resistance to the flow of current, which is indicative of its antiparallel (AP) state [41,42].

The resistance of antiparallel or parallel states decides the logic '0' or high resistance state (HRS) and the logic '1' or low resistance state (LRS) of the device [43]. In the case of MTJ, due to the presence of an insulating layer, the tunnel magnetoresistance (TMR) effect is observed. The TMR ratio is an important parameter for MTJ and is given as follows [22]:

$$\text{Tunnel magnetoresistance (TMR)} = \frac{R_{AP} - R_P}{R_P}$$

where R_P and R_{AP} denote the value of resistance in parallel and anti-parallel configuration in MTJ, respectively. Therefore, MTJ can be considered a two-valued resistor, which is very significant in memory and logical applications. MTJ is assumed to have stored bits '1' and '0' when it is in P and AP configurations, respectively, and the stored data values are nonvolatile [40].

7.5   MAGNETIC TUNNEL JUNCTION (MTJ) WRITING TECHNIQUES

The writing process is also known as switching or storing the data in MTJ from parallel to anti-parallel configuration or vice versa and can be achieved by switching the magnetic orientation of the fixed layer (FL). The various mechanisms used for MTJ writing process are discussed below.

7.6   FIELD-INDUCED MAGNETIC SWITCHING (FIMS)

Field-induced magnetic switching (FIMS) is the first-generation MRAM write technique that uses an external magnetic field to change the magnetic orientation of fixed layer in MTJs [44]. A current is passed through orthogonal write lines (also called bit and digit lines) to generate the magnetic field, as shown in Figure 7.6.

As orthogonal currents flow in bit lines and digit lines, respectively, they produce hard-axis and easy-axis switching fields, which turn on the MTJs. Depending on the flow of the current direction, parallel (P) and anti-parallel (AP) configurations are switched. Despite its advantages during sensing, this writing technique has a number of disadvantages, such as low density, high power consumption, and limited scalability [44].

The current needed to produce a magnetic field for MTJ writing is high ($\gtrsim 10mA$), as a result the power consumption increases. Furthermore, the electromigration effect limits scalability, which makes it difficult to reach low densities. MTJs are arranged in an array in which the magnetic orientation of FL is toggled by the external field generated. This results in erroneous switching of other neighboring MTJs in that array. To alleviate this problem, a novel toggle switching mode was proposed by Freescale. Based on this writing technique, they launched the first commercial 4 Mb MRAM product [45]. Though the toggle switching approach addressed the concern of the half-selectivity problem, it did not resolve the issue of high power consumption, low density, and limited scalability.

7.7   THERMAL-ASSISTED SWITCHING (TAS)

Thermal-assisted switching (TAS) is an upgraded writing technique compared to field-induced magnetic switching (FIMS) [46]. Figure 7.7 shows the writing process

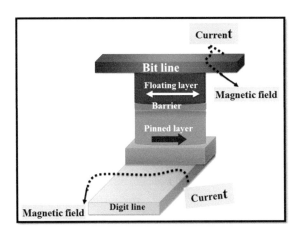

FIGURE 7.6   The writing process in field-induced magnetic switching (FIMS).

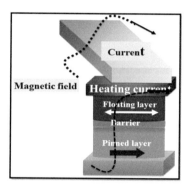

FIGURE 7.7   The writing process in thermal-assisted switching (TAS).

in TAS. In the case of TAS, a heating current is passed through the MTJ to be written, which lowers the magnetic field strength during writing. Hence the generation of magnetic fields can be achieved by a single current line and as a result the writing energy and circuit area are reduced significantly. As a result of its power savings, higher density, and higher thermal stability, TAS is utilized in MRAM [47,48] and look-up tables (LUTs) [49]. TAS, however, has been limited in its application to high-speed logic applications due to the requirement for a relatively longer cooling duration after being switched.

7.8 CONCLUSION

An overview of spintronic devices, which are expected to become a mainstream technology in the future, is presented in this chapter. This chapter particularly focused on different types of spin-valve devices and their working principles, along with different conventional and recently developed types of MTJ devices. The most important characteristics of the spintronic devices, such as GMR and TMR effects, were discussed. The various writing techniques adapted for MTJs such as FIMS and TAS, and their switching dynamics, were outlined. Among the different types of spintronic devices, MRAM-based memory devices are dominant and play an important in memory technology. Several spintronic companies have succeeded in commercializing embedded MRAM memory products. In addition to memory applications, spintronics are also expected to be used in the recently buzzed-about PIM paradigm of normally-off/instant-on on computer architecture. The CMOS technology is currently being integrated with spintronic devices using CAD tools such as Cadence in order to develop applications for memory and hybrid circuits. However, the simulation models and prototypes differ substantially. Moreover, in order to increase the speed and lower the threshold current at which data bits can be manipulated, artificially engineered thin-film structures are needed. The field of spintronics is still in its infancy and requires much more research, both academically and industrially, in order to fill this void. The aim is to develop low-power, high-speed, high-density circuits by integrating microelectronics with new materials, fabrication methods, and new computer architectures.

REFERENCES

[1] Kim, N.S., Austin, T., Blaauw, D., Mudge, T., Hu, J.S., Irwin, M.J., Kandemir, M., Narayanan, V., et al. (2003) Leakage current: Moore's law meets static power. *Computer* 36(12): 68–75.

[2] Gariglio, S. (2020) Electric control of a spin current has potential for low-power computing. *Nature* 580: 458–459.

[3] Transistor count – Wikipedia (2020) [online; accessed Jul 2020].

[4] Waldrop, M.M. (2016) The chips are down for Moore's law. *Nature News* 530: 144.

[5] Lin, X., Yang, W., Wang, K.L., Zhao, W. (2019) Two-dimensional spintronics for low-power electronics. *Nature Electronics* 2(7): 274–283.

[6] Liu, W., Wong, P.K.J., Xu, Y. (2019) Hybrid spintronic materials: growth, structure and properties. *Progress in Materials Science* 99: 27–105.

[7] Joshi, V.K. (2016) Spintronics: a contemporary review of emerging electronics devices. *Engineering Science and Technology* 19(3): 1503–1513.

[8] Yakout, S.M. (2020) Spintronics: future technology for new data storage and communication devices. *Journal of Superconductivity and Novel Magnetism* 33: 2557–2580.

[9] Chang, C.-Y. (2003) The highlights in the nano world. *Proceedings of the IEEE* 91(11): 1756–1764.

[10] Dieny, B., Prejbeanu, I.L., Garello, K., Gambardella, P., Freitas, P., Lehndorff, R., Raberg, W., Ebels, U., Demokritov, S.O., Akerman, J., Deac, A., Pirro, P., Adelmann, C., Anane, A., Chumak, A.V., Hirohata, A., Mangin, S., Valenzuela, S.O., Onbaşlı, M.C., D'Aquino, M., Prenat, G., Finocchio, G., Lopez-Diaz, L., Chantrell, R., Chubykalo-Fesenko, O., Bortolotti, P. (2020) Opportunities and challenges for spintronics in the microelectronics industry. *Nature Electronics* 3: 446–459.

[11] Puebla, J., Kim, J., Kondou, K., Otani, Y. (2020) Spintronic devices for energy-efficient data storage and energy harvesting. *Communications Materials* 1(24): 1–9.

[12] Vedmedenko, E.Y., Kawakami, R.K., Sheka, D.D., Gambardella, P., Kirilyuk, A., Hirohata, A., Binek, C., Chubykalo-Fesenko, O., Sanvito, S., Kirby, B.J., Grollier, J., Everschor-Sitte, K., Kampfrath, T., You, C.-Y., Berger, A. (2020) The 2020 magnetism roadmap. *Journal of Physics D: Applied Physics* 53(45): 453001.

[13] Joshi, V.K., Barla, P., Bhat, S., Kaushik, B.K. (2020) From MTJ device to hybrid CMOS/MTJ circuits: a review. *IEEE Access* 8: 194105–194146.

[14] Tsang, C., Fontana, R.E., Lin, T., Heim, D.E., Speriosu, V.S., Gurney, B.A., Williams, M.L. (1994) Design, fabrication and testing of spin-valve read heads for high density recording. *IEEE Transactions on Magnetics* 30(6): 3801–3806.

[15] Bandyopadhyay, S., Cahay, M. (2015) *Introduction to Spintronics*. CRC Press, Boca Raton.

[16] Zhao, W., Prenat, G. (2015) *Spintronics-Based Computing*. Springer International Publishing, Berlin.

[17] Wolf, S.A., Awschalom, D.D., Buhrman, R.A., Daughton, J.M., von Molnár, S., Roukes, M.L., Chtchelkanova, A.Y., Treger, D.M. (2001) Spintronics: a spin-based electronics vision for the future. *Science* 294(5546): 1488–1495.

[18] Rasool, A., Amiruddin, R., Mohamed, I.R. and Kumar, M.S. (2020) Fabrication and characterization of resistive random access memory (ReRAM) devices using molybdenum trioxide (MoO3) as switching layer. *Superlattices and Microstructures*, 147: 106682..

[19] Žutić, I., Fabian, J., Das Sarma, S. (2004) Spintronics: fundamentals and applications. *Reviews of Modern Physics* 76(2): 323–410.

[20] Endoh, T., Koike, H., Ikeda, S., Hanyu, T., Ohno, H. (2016) An overview of nonvolatile emerging memories— spintronics for working memories. *IEEE Journal on Emerging and Selected Topics in Circuits and Systems* 6(2): 109–119.

[21] Kang, W., Deng, E., Wang, Z., Zhao, W. (2020) Spintronic logic-in memory paradigms and implementations. In: Suri, M. (ed.) *Applications of Emerging Memory Technology. Springer Series in Advanced Microelectronics*, 63: 215–229.

[22] Maurice, D.P.A., Howard, F.R. (1928) The quantum theory of the electron. *Proceedings of the Royal Society of London* A 117: 610–624.

[23] Roup, R.R., Kilby, J.S. (1958) *Electrical circuit elements*. US Patent 2,841,508.

[24] Kilby, J.S. (1976) Invention of the integrated circuit. *IEEE Transactions on Electron Devices* 23(7): 648–654.

[25] Baibich, M.N., Broto, J.M., Fert, A., Van Dau, F.N., Petroff, F., Etienne, P., Creuzet, G., Friederich, A., Chazelas, J. (1988) Giant magnetoresistance of (001)Fe/(001)Cr magnetic superlattices. *Physical Review Letters* 61(21): 2472–2475.

[26] Binasch, G., Grünberg, P., Saurenbach, F., Zinn, W. (1989) Enhanced magnetoresistance in layered magnetic structures with antiferromagnetic interlayer exchange. *Physical Review B* 39: 4828–4830.

[27] Ennen, I., Kappe, D., Rempel, T., Glenske, C. and Hütten, A. (2016) Giant magnetoresistance: basic concepts, microstructure, magnetic interactions and applications. *Sensors* 16(6): 904.

[28] Datta, S., Das, B. (1990) Electronic analog of the electro-optic modulator. *Applied Physics Letters* 56(7): 665–667.

[29] Shinjo, T. (2013) *Nanomagnetism and Spintronics*. Elsevier, Amsterdam.

[30] Reig, C., Cubells-Beltrán, M.-D., Ramírez Muñoz, D. (2009) Magnetic field sensors based on giant magnetoresistance (GMR) technology: applications in electrical current sensing. *Sensors* 9(10): 7919–7942.

[31] Dieny, B., Speriosu, V.S., Parkin, S.S.P., Gurney, B.A., Wilhoit, D.R., Mauri, D. (1991) Giant magnetoresistive in soft ferromagnetic multilayers. *Physics Review B* 43(1): 1297–1300(R).

[32] Feng, Y.P., Shen, L., Yang, M., Wang, A., Zeng, M., Wu, Q., Chintalapati, S., Chang, C.-R. (2017) Prospects of spintronics based on 2D materials. *WIREs Computational Molecular Science* 7(5): e1313.

[33] Fong, X., Kim, Y., Yogendra, K., Fan, D., Sengupta, A., Raghunathan, A., Roy, K. (2015) Spin-transfer torque devices for logic and memory: prospects and perspectives. *IEEE Transactions on Computer-Aided Design of Integrated Circuits and Systems* 35(1): 1–22.

[34] Ji, Y., Hoffmann, A., Jiang, J.S., Bader, S.D. (2004) Spin injection, diffusion, and detection in lateral spin-valves. *Applied Physics Letters* 85(25): 6218–6220.

[35] Fukuma, Y., Wang, L., Idzuchi, H., Takahashi, S., Maekawa, S., Otani, Y. (2011) Giant enhancement of spin accumulation and long-distance spin precession in metallic lateral spin valves. *Nature Materials* 10(7): 527–531.

[36] Engel, B.N., Rizzo, N.D., Janesky, J., Slaughter, J.M., Dave, R., DeHerrera, M., Durlam, M., Tehrani, S. (2002) The science and technology of magnetoresistive tunneling memory. *IEEE Transactions on Nanotechnology* 99(1): 32–38.

[37] Ho, M.K., Tsang, C.H., Fontana Jr., R.E., Parkin, S.S.P., Carey, K.J., Pan, T. MacDonald, S., Arnett, P.C., Moore, J.O. (2001) Study of magnetic tunnel junction read sensors. *IEEE Transactions on Magnetics* 37(4): 1691–1694.

[38] Zhu, X., Zhu, J.-G. (2007) Domain wall pinning and corresponding energy barrier in percolated perpendicular medium. *IEEE Transactions on Magnetics* 43(6): 2349–2353.

[39] Kaisha Toshiba, K. (2012) *Random number generator.* US Patent, 2012/0026, 784, 2012-02-22.

[40] Dijken, S.V., Jiang, C., Parkin, S.S. (2002) Room temperature operation of a high output current magnetic tunnel transistor. *Applied Physics Letters* 80(18): 3364–3366.

[41] Shuto, Y., Nakane, R., Wang, W., Sukegawa, H., Yamamoto, S., Tanaka, M., Inomata, K., Sugahara, S. (2010) A new spin-functional metal–oxide–semiconductor field-effect transistor based on magnetic tunnel junction technology: pseudo-spin-MOSFET. *Applied Physics Express* 3(1): 013003-1-3.

[42] Krzysteczko, P., Reiss, G., Thomas, A. (1990) Memristive switching of MgO based magnetic tunnel junctions. Applied Physics Letters 95(11): 112508-1-3.

[43] Singh, J.P., Kaur, B., Gautam, S., Lim, W.C., Asokan, K., Chae, K.H. (2016) Chemical effects at the interfaces of Fe/MgO/Fe magnetic tunnel junction. *Superlattices and Microstructures* 100: 560–586.

[44] Wolf, S.A., Awschalom, D.D., Buhrman, R.A., Daughton, J.M., von Molnár, S., Roukes, M.L., Chtchelkanova, A.Y., Treger, D.M. Spintronics: a spin-based electronics vision for the future. *Science* 294(5546): 1488–1495.

[45] Engel, B.N., Akerman, J., Butcher, B., Dave, R.W., DeHerrera, M., Durlam, M., Grynkewich, G., Janesky, J., Pietambaram, S.V., Rizzo, N.D., Slaughter, J.M., Smith, K., Sun, J.J., Tehrani, S. (2005) A 4-Mb toggle MRAM based on a novel bit and switching method. *IEEE Transactions On Magnetics* 41(1): 132–136.

[46] Prejbeanu, I.L., Kula, W., Ounadjela, K., Sousa, R.C., Redon, O., Dieny, B., Nozieres, J.-P. (2004) Thermally assisted switching in exchange-biased storage layer magnetic tunnel junctions. *IEEE Transactions on Magnetics* 40(4): 2625–2627.

[47] Prejbeanu, I.L., Kerekes, M., Sousa, R.C., Sibuet, H., Redon, O., Dieny, B., Nozières, J.P. (2007) Thermally assisted MRAM. *Journal of Physics: Condensed Matter* 19(16): 165218.

[48] Prejbeanu, I.L., Bandiera, S., Alvarez-Hérault, J., Sousa, R.C., Dieny, B., Nozières, J.-P. (2013) Thermally assisted MRAMs: ultimate scalability and logic functionalities. *Journal of Physics D: Applied Physics* 46(7): 074002.

[49] Zhao, W., Belhaire, E., Chappert, C., Dieny, B., Prenat, G. (2009) Tasmram-based low-power high-speed runtime reconfiguration (rtr) fpga. *ACM Transactions on. Reconfigurable Technology and Systems (TRETS)* 2(2): 8.

8 Mathematical Approach for a Future Semiconductor Roadmap

Shiromani Balmukund Rahi[1], Abhishek Kumar Upadhyay[2], Young Suh Song[3], Nidhi Sahni[4], Ramakant Yadav[5], Umesh Chandra Bind[6], Guenifi Naima[7], Billel Smaani[8], Chandan Kumar Pandey[9], Samir Labiod[10], T.S. Arun Samul[11], Hanumant Lal[12], H. Bijo Josheph[13]

[1]Department of Electrical Engineering, Indian Institute of Technology, Kanpur 208016, India
[2]X-FAB Semiconductor foundries, 99097 Erfurt, Germany
[3]Korea Military Academy, Seoul, Republic of Korea
[4]Department of Mathematics, Sharda School of Basic Sciences and Research, Sharda University Greater Noida, Uttar Pradesh-201310, India
[5]Department of Electrical & Electronics Engineering, Mahindra University, Hyderabad 500043, India
[6]Centre of Nanotechnology, Indian Institute of Technology Roorkee, Roorkee 247667, India
[7]LEA Electronics Department, University Mostefa Benboulaid of Batna 2, Batna - 05000, Algeria
[8]Centre Universitaire Abdelhafid Boussouf – Mila, 43000, Algeria
[9]VIT-AP University, Amaravati, Andhra Pradesh India
[10]Laboratoire d'Automatique Appliquée, Université M'Hamed Bougara de Boumerdes, Algeria
[11]Department of ECE, National Engineering College, Kovilpatti, 628503, India
[12]Department of Electronics and Communication Engineering, Buddha Institute of Technology, Gorakhpur India
[13]Department of Physics and Nanotechnology, SRM Institute of Science and Technology Channi, India

DOI: 10.1201/9781003373391-8

8.1 INTRODUCTION

Conventional MOS transistors: The progress of the very large-scale integration (VLSI) world has been possible with continuous scaling of silicon-based metal–oxide–semiconductor field effect transistors (MOSFETs). This device has been constantly scaled down in pursuit of lower power consumption, enhanced performance, smaller area, and lower cost. Aggressive scaling is a cause of various limitations and challenges in classical MOSFET. However, it has been expected that the down-scaling will reach its limits about a gate length (L_G) of 5 nm [1–4]. Leakage current is one of the most important causes of the scaling limitation of classical Si-based MOSFET. The leakage current ($I_{Leakage}$) available in conventional MOS devices is one of the most challenging factors; especially for low-power energy (LPE)-based applications. Especially, existing leakage current components are a major road blocker in the path of ultra-low-power (ULP) circuit and system developments. Ultra-low-voltage operation of CMOS circuits significantly reduces the circuits' power consumption. Recently, reducing energy consumption has been the most common practice worldwide. For the purpose of intelligently lowering power consumption of electronic equipment, semiconductor devices, including CMOS LSIs, clearly need to enhance their power efficiency significantly. In order to efficiently lower the power consumption of CMOS circuits, lowering the supply voltage (V_{dd}) is most effective. Figure 8.1 shows the ITRS prediction for the possible development and scaling trend. In Figure 8.1, there are three sub-windows showing the trends of (a) speed, (b) energy, and (c) gate length in HP devices in the past (lEDM) and future (2011 ITRS). Lowering the gate length indicates continuous scaling of MOSFETs.

Subthreshold swing in semiconductors is a crucial parameter, primarily observed in metal-oxide semiconductor field effect transistors (MOSFETs) and similar transistor devices. It represents the energy efficiency required for the transistor to transition between the ON and OFF states by controlling the applied voltage.

Subthreshold swing plays a significant role in devices and sources that consume low power or operate on limited resources like batteries. A smaller subthreshold swing enables transistors to operate more efficiently, consuming less power even with small input signals. This advancement contributes to the progress of semiconductor technology, leading to improved performance in low-power applications such as mobile devices.

Nevertheless, improving subthreshold swing is not an easy task due to various physical limitations arising from existing device structures and materials. To address this challenge, extensive research and development into novel transistor designs and materials are required in the semiconductor industry. These efforts hold the promise of enhancing power efficiency and overall performance in semiconductor devices.

This steady and consistent reduction, however, has been extremely difficult for conventional bulk CMOS because of increasing variability and decreasing voltage margin between V_{dd} and threshold voltage (V_{th}) [5]. Figure 8.1(d) shows that the inverse subthreshold slope (SS) has an increasing trend with scaling in FinFET. Therefore, increasing behavior of SS suggests that short-channel effects (SCEs) will be more prominent at the lower contracted-gate-pitch. As shown in Figure 8.1(d), for sub-20 nm channel lengths, DIBL increases; therefore, the FinFET devices are approaching the short-channel regime.

The basic requirements of circuit and system design are formulated as 'figure of merit' (FoM) [6]:

$$Figure\, of\, Merit = \frac{Intelligence}{\left(\left(Size\right)\times\left(Cost\right)\times\left(Power\right)\right)}$$

The theoretical measurement greatly depends on scaling of the existing FET devices. The bulk MOSFETs have played a lead role in the development of VLSI circuits and systems in the form of integrated circuits (ICs). Existing various SOCs

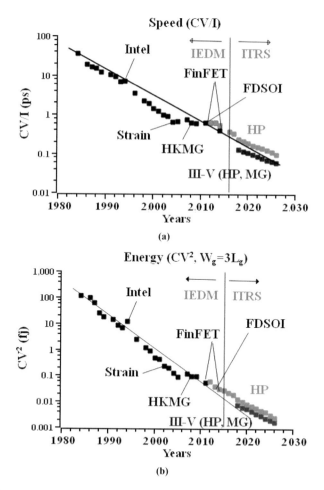

FIGURE 8.1   The trends of (a) speed, (b) energy, (c) gate length in HP devices in the past (IEDM) and future (2011 ITRS), and (d) gate length.

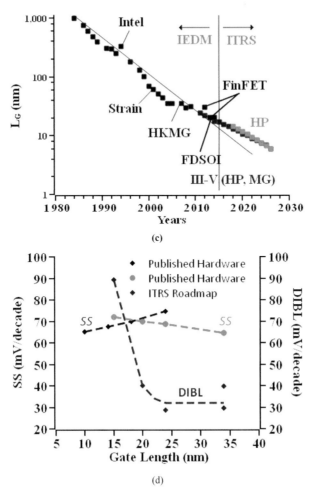

FIGURE 8.1   (Continued)

(system on chips) are examples of such types of development. Reduction of power in modern technology directly realizes an improvement in the device and circuit as well as the system based on MOS devices. However, the power factor is mainly affected by existing leakage power. In conventional MOSFET, there are various components of leakage, as shown in Figure 8.2. The existing leakages in MOSFET include: (1) the reverse-bias PN junction leakage denoted as (I_1), (2): subthreshold leakage (I_2), (3) the oxide tunneling current (I_3), (4) the gate current due to hot-carrier injection (I_4), (5) the GIDL (I_5), and (6) the channel punch through current (I_6). The leakage current components (2), (5), and (6) are off-state leakage mechanisms, while (1) and (3) occur in both ON and OFF states, and (4) occurs in the off state [6].

   In Equations (8.1) and (8.2) the used terms, V_{th}, V_T, C_{OX}, μ_0, $\acute{\eta}$, T_D, C_D, and T_{OX} are the threshold voltage, thermal voltage, gate oxide capacitance, zero bias mobility, body effect coefficient, maximum depletion layer width, depletion layer capacitance, and thickness of the gate oxide, respectively.

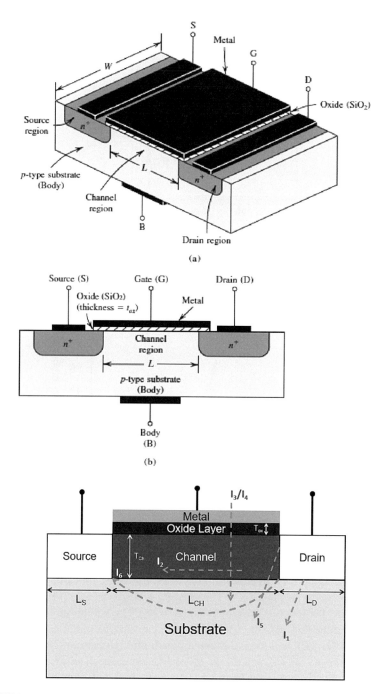

FIGURE 8.2 Conventional scaling and its impact on dc characteristics.

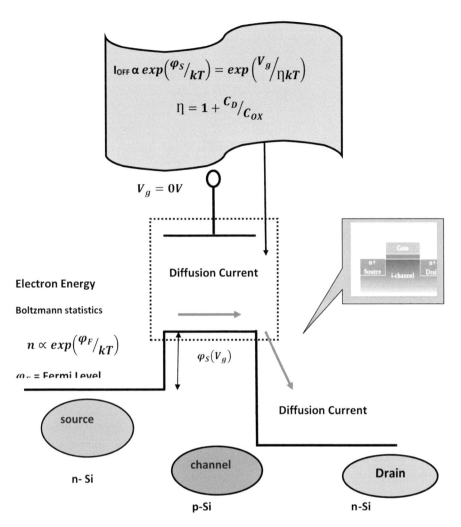

$$I_{OFF} \, \alpha \, exp\left(\varphi_S/kT\right) = exp\left(V_g/\eta kT\right)$$

$$\eta = 1 + C_D/C_{OX}$$

$$V_g = 0V$$

Diffusion Current

Electron Energy

Boltzmann statistics

$$n \propto exp\left(\varphi_F/kT\right)$$

φ_F = Fermi Level

$\varphi_S(V_g)$

Diffusion Current

source

n- Si

channel

p-Si

Drain

n-Si

FIGURE 8.3 Sketch diagram of leakage current transport phenomena in conventional MOSFET.

A generalized sketch diagram for leakage current in conventional MOS devices is shown in Figure 8.3 Another, more important low-power (LP) device and circuit design parameter in the subthreshold region is the inverse-subthreshold slope (SS), which indicates the speed of turning off the transistor below the threshold voltage and is given by Equation (8.3) [7–13]. The used variable in SS for conventional MOSFET has a common meaning like k, Boltzmann constant, having a value = 1.381E-23 J•K–1. T is temperature in Kelvin. The elementary charge is denoted by q which has a numerical value of 1.602176634×10^{-19} coulombs.

The mathematical expression of the subthreshold leakage current is as follows [7–13]:

$$I_{DS} = \mu_0 C_{OX} \frac{W}{L} (\eta - 1) V_T^2 e^{\frac{V_G - V_{TH}}{\eta V_T}} (1 - e^{V_{DS}/V_T}) \tag{8.1}$$

$$\eta = 1 + \frac{C_D}{C_{OX}} = 1 + \frac{3T_{OX}}{T_D} \tag{8.2}$$

$$SS = \left(\frac{d(\log_{10} I_{DS})}{dV_{GS}} \right)^{-1} = 2.3 \frac{kT}{q} \left(1 + \frac{C_D}{C_{OX}} \right) = 2.3\eta \frac{kT}{q} \tag{8.3}$$

The SS defined for conventional MOS devices is defined by Equation (8.3), which is the change in gate voltage (VGS) needed to increase the drain current (IDS) by one order of magnitude. Measured in units of millivolts per decade, in conventional MOSFETs it is limited to kT/q by the Boltzmann electron energy distribution [6–12]. The numerical value of Boltzmann electron energy at room temperature ($T = 300K$) is 26 mV. At room temperature, the numerical value of SS is 60 mV/decade. The lower the value of SS, the better is the MOS device. As suggested by the expression, the lower value can be obtained by (a) reducing temperature T, (b) higher C_{ox} or a thinner oxide layer, and (c) lower C_{dm} or a thicker depletion layer or lower substrate doping [13–19].

8.2 POWER FACTOR WITH SCALED MOS TRANSISTORS OPPORTUNITY

The power dissipation of the CMOS circuit and system is roughly measured by Equation (8.4), having two components, dynamic power and static power. The dynamic power is measure by mathematical relation (8.5) and has strong dependency of power supply V_{DD}. In the case of conventional CMOS VLSI circuits and systems, lowering the power supply V_{DD} becomes a challenging task. Figure 8.4 shows the trend of power supply with time and future with negative capacitance MOS technology

$$P_{dissipation} = Dynamicpower + Staticpower \tag{8.4}$$

$$Dynamicpower = \alpha C_{load} \times V_{DD}^2 \tag{8.5}$$

$$Staticpower = V_{DD} \left(I_{leakage} + I_{th} \times 10^{-\left(V_T/SS \right)} \right) \tag{8.6}$$

In Equations (8.5) and (8.6), α is the activity factor, f the operating frequency, and SS is subthreshold slope; $I_{leakage}$ shows the summation of leakage current from gate, the junctions, and the band-to-band tunneling (BTBT), while I_{th} is the drain current at V_T. For the purpose of intellectually maintaining low power consumption, lower V_{DD}, and leakage current, higher V_T and steeper SS are required. In any logic MOSFET, a large-enough ratio between the drain current I_d in the on-state and in the off-state must

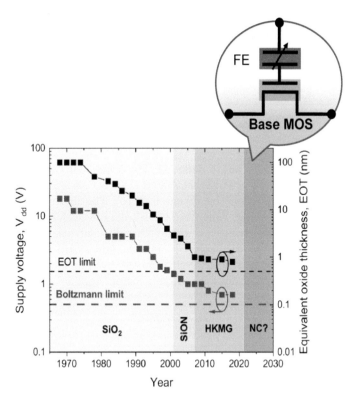

FIGURE 8.4 Scaling trend of power supply with time and in the future with negative capacitance MOS technology.

be ensured to accomplish the desirable and acceptable performance (high I_{on}) and off-state (also known as static) power consumption (low I_{off}) at the same time [14–26].

The rapid scaling of conventional FET device design parameters maintains performance and these requirements cause an exponential increase in leakage current (I_{OFF}). Equations (8.7), (8.8), and (8.9) are alternatives for the power dissipation of the VLSI circuit and system. With the help of these equations, the importance of power scaling, V_{DD}, can be seen.

$$P_{total} = P_{static} + P_{dynamic} \tag{8.7}$$

$$P_{static} = N_g \times I_{off} \times V_{DD} \tag{8.8}$$

$$P_{dynamic} = \alpha \times C_{totl} \times V_{DD}^2 \times f \tag{8.9}$$

Ng is the number of gates, α is the fraction of active gates, C_{total} the total load capacitance of all gates, and f switching frequency. Figure 8.1 illustrates the leakage current in conventional MOSFET and its dependency on the device design matrix component. In-depth and detailed investigations regarding these are beyond the scope of this chapter. However, readers can refer to suitable articles and books. Figure 8.3 presents the power supply scaling over time. Figure 8.3 shows that negative capacitance FETs will play a lead role in the future [27–38].

8.3 SCOPE AND LIMITATIONS OF CONVENTIONAL CMOS TECHNOLOGY

Figure 8.5 shows the scope and limitations of conventional CMOS technology. Semiconductor devices are generally classified into two categories based on the subthreshold slope (SS) defined by Equation (8.10). The first category contains MOS devices having SS values larger than 60 mV/decade, such as bulk MOSFET, FinFET, FDSOI, NWFET, nanosheet FET (NSFET), and gate all around FET (GAA). In expression 'SS', the factor 'n' in Equation (8.10) for conventional MOSFET could be mathematically reinterpreted by Equation (8.11), which is known as equation describing the transport factor. Another category of MOS

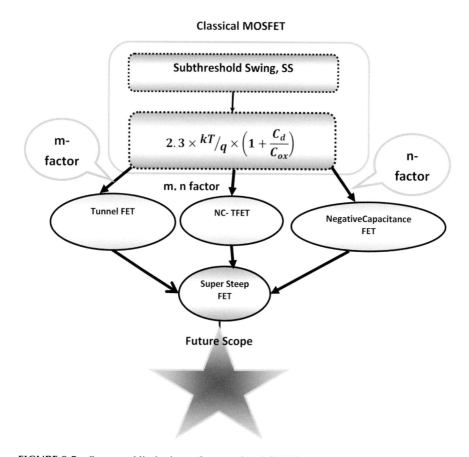

FIGURE 8.5   Scope and limitations of conventional CMOS technology.

devices has SS lower than 60 mV/decade at room temperature, known as steep FET. The most popular candidates in this category are tunnel FET and NCFET. In the case of tunnel FET, charge transport is governed by band-to-band tunneling in reverse-based operation conditions. This category of devices, like double gate tunnel FET, high-K tunnel FET, GAA tunnel FET, and other developments have the desirable lower off-state current. However, due to band-to-band tunneling, on-state charge transport is limited. This causes a lower value of on-state current than for conventional MOSFET. This is because of lower switching than conventional CMOS devices. This is a specific drawback with the tunnel FET technology. As a rule of thumb, a measurement is conducted by semiconductor engineers. However, there remain several issues to overcome to realize the continuous improvement of semiconductor technology [33–43].

$$n = \frac{d\Psi_S}{d\left(log_{10} I_{DS}\right)} = log\left(10\right) \times \frac{k_B T}{q} \qquad (8.10)$$

$$= 2.3 \times 25.8 \frac{mV}{decade} \approx 60 \frac{mV}{decade} \qquad (8.11)$$

Figure 8.5 illustrates the opportunity for research and development with steep subthreshold swing FET devices, circuits, and systems. The availability of negative capacitance in ferroelectric materials requires new research and development with experts on ferroelectric or negative capacitance [19–50].

8.4 CONCLUSION

In this chapter, we have carefully investigated the key purpose of designing modern semiconductors, from various perspectives. Especially, the history of the modern semiconductor industry has been covered so that the key factors in future semiconductor design can be easily and intuitively understood. Low subthreshold swing transistor design is the most important design methodology, since it can achieve low power consumption and high performance of transistors at the same time.

8.5 ACKNOWLEDGMENT

This chapter has been written by the corresponding author (Dr. Shiromani Balmukund Rahi). The author has written and edited this chapter by referring to various earlier research. The author greatly appreciates the relevant earlier research which helped the author in writing this manuscript.

REFERENCES

1. Iwai, H. (2008) CMOS technology after reaching the scale limit. In: *Extended Abstracts – 2008 8th International Workshop on Junction Technology (IWJT '08)*, pp. 1–2.

2. Lundstrom, M. (2003) Device physics at the scaling limit: what matters? [MOSFETs]. In: *IEEE International Electron Devices Meeting 2003*, pp. 33.1.1–33.1.4.
3. Anas, M., Amin, S.I., Beg, M.T. et al. (2022) Design and analysis of GaSb/Si based negative capacitance TFET at the device and circuit level. *Silicon*.14: 11951–11961.
4. Roy, K., Mukhopadhyay, S. and Mahmoodi-Meimand, H. (2003) Leakage current mechanisms and leakage reduction techniques in deep-submicrometric CMOS circuits. *Proceedings of the IEEE* 91(2): 305–327.
5. Hiramoto, T. et al. (2016) Ultra-low power and ultra-low voltage devices and circuits for IoT applications. In: *2016 IEEE Silicon Nanoelectronics Workshop (SNW)*, pp. 146–147.
6. Makimoto, T. and Sakai, Y. (2003) Evolution of low power electronics and its future applications. In: *Proceedings of the 2003 International Symposium on Low Power Electronics and Design, 2003. ISLPED '03*, pp. 2–5.
7. Taur, Y. and Ning, T. H. (1998) *Fundamentals of Modern VLSI Devices*. New York: Cambridge University Press, ch. 2, pp. 94–95.
8. Roy, K., Mukhopadhyay, S. and Mahmoodi-Meimand, H. (2003) Leakage current mechanisms and leakage reduction techniques in deep-submicrometric CMOS circuits. *Proceedings of the IEEE* 91(2): 305–327.
9. Pierret, R. (1996) *Semiconductor Device Fundamentals*. Reading, MA: Addison-Wesley, ch. 6, pp. 235–300.
10. Taur, Y. and Ning, T. H. (1998) *Fundamentals of Modern VLSI Devices*. New York: Cambridge University Press, ch. 2, pp. 94–95.
11. Rabaey, J. M. (1996) *Digital Integrated Circuits*. Englewood Cliffs, NJ: Prentice-Hall, ch. 2, pp. 55–56.
12. Taur Y. and Ning, T. H. (1998) *Fundamentals of Modern VLSI Devices*. New York: Cambridge University Press, ch. 3, pp.143–144.
13. Paul, B. C. Soeleman, H. and Roy, K. (2001) An 8 × 8 subthreshold digital CMOS carry save array multiplier In: *Proceedings of the 27th European Solid-State Circuits Conference (ESSCIRC '01)*, pp. 377–380, Villach, Austria, September 2001.
14. Singh, K.J., Bulusu, A. and Dasgupta, S. (2022) Origin of negative capacitance transient in ultrascaled multidomain metal-ferroelectric-metal stack and hysteresis-free Landau transistor. *IEEE Transactions on Electron Devices* 69(3): 1284–1292.
15. Wei, C. and Banerjee, K. (2020) Is negative capacitance FET a steep-slope logic switch? *Nature Communications* 11.1: 1–8.
16. A. Saeidi et al. (2017) Negative capacitance as performance booster for tunnel FETs and MOSFETs: An experimental study. *IEEE Electron Device Letters* 38(10): 1485–1488.
17. Kobayashi, M., Jang, K., Ueyama, N. and Hiramoto, T. (2016) Negative capacitance as a performance booster for tunnel FET. In: *2016 IEEE Silicon Nanoelectronics Workshop (SNW)*, pp. 150–151.
18. Kim, H.W. and Kwon, D. (2021) Gate-normal negative capacitance tunnel field-effect transistor (TFET) with channel doping engineering. *IEEE Transactions on Nanotechnology* 20: 278–281.
19. Lin, C.-I., et al. (2016) Effects of the variation of ferroelectric properties on negative capacitance FET characteristics. *IEEE Transactions on Electron Devices* 63(5): 2197–2199.
20. Cao, W. and Banerjee, K. Is negative capacitance FET a steep-slope logic switch? *Nature Communications* 11(1): 1–8.
21. Salahuddin, S. and Datta, S. (2008) Use of negative capacitance to provide voltage amplification for low power nanoscale devices. *Nano Letters* 8(8): 405–410.

22. Zhirnov, V. V. and Cavin, R. K. Negative capacitance to the rescue? *Nature Nanotechnology* 3(2): 77–78.

23. Khan, A. I., et al. (2011) Ferroelectric negative capacitance MOSFET: Capacitance tuning & antiferroelectric operation." In: *2011 International Electron Devices Meeting*. IEEE.

24. Catalan, G., Jiménez, D. and Gruverman, A. Negative capacitance detected. *Nature Materials* 14(2): 137–139.

25. Rahi, S. B., Tayal, S. and Upadhyay, A. K. (2021) A review on emerging negative capacitance field effect transistor for low power electronics. *Microelectronics Journal* 116: 105242.

26. Upadhyay, A. K., Rahi, S. B., Tayal, S. and Song, Y. S. (2022) Recent progress on negative capacitance tunnel FET for low-power applications: Device perspective. *Microelectronics Journal* 2022: 105583.

27. Amrouch, H., van Santen, V. M., Pahwa, G., Chauhan, Y. and Henkel, J. NCFET to rescue technology scaling: Opportunities and challenges. In: *2020 25th Asia and South Pacific Design Automation Conference (ASP-DAC)*, pp. 637–644.

28. You, W.-X., Su, P. and Hu, C. (2018) Evaluation of NC-FinFET based subsystem-level logic circuits using SPICE simulation. In: *2018 IEEE SOI-3D-Subthreshold Microelectronics Technology Unified Conference (S3S)*, pp. 1–2.

29. Sakib, F. I., Hasan, M. A. and Hossain, M. (2020) Exploration of negative capacitance in gate-all-around Si nanosheet transistors. *IEEE Transactions on Electron Devices* 67(11): 5236–5242.

30. Kim, M., Seo, J. and Shin, M. (2018) Biaxial strain based performance modulation of negative-capacitance FETs. In: *2018 International Conference on Simulation of Semiconductor Processes and Devices (SISPAD)*, pp. 318–322.

31. Hoffmann, M., Slesazeck, S., and Mikolajick, T. (2021) Progress and future prospects of negative capacitance electronics: A materials perspective. *APL Materials* 9(2): 020902.

32. Dong, Z. and Guo, J. (2017) A simple model of negative capacitance FET with electrostatic short channel effects. *IEEE Transactions on Electron Devices* 64(7): 2927–2934.

33. Li, X., et al. Enabling energy-efficient nonvolatile computing with negative capacitance FET. *IEEE Transactions on Electron Devices* 64(8): 3452–3458

34. Kwon, D., et al. Negative capacitance FET with 1.8-nm-thick Zr-doped HfO_2 oxide. *IEEE Electron Device Letters* 40(6): 993–996.

35. Li, K.-S., et al. Sub-60mV-swing negative-capacitance FinFET without hysteresis. In: *2015 IEEE International Electron Devices Meeting (IEDM)*. IEEE, 2015.

36. Kobayashi, M., et al. Experimental study on polarization-limited operation speed of negative capacitance FET with ferroelectric HfO_2. In: *2016 IEEE International Electron Devices Meeting (IEDM)*. IEEE, 2016.

37. Tayal, S., Rahi, S. B., Srivastava, J. P. and Bhattacharya, S. (2021) Recent trends in compact modeling of negative capacitance field-effect transistors." In: *Semiconductor Devices and Technologies for Future Ultra Low Power Electronics*, pp. 203–226. CRC Press.

38. Tayal, S., Upadhyay, A. K., Kumar, D. and Rahi, S. B. (eds.) (2022) *Emerging Low-Power Semiconductor Devices: Applications for Future Technology Nodes*. CRC Press.

39. Hu, V. P.-H., et al. Optimization of negative-capacitance vertical-tunnel FET (NCVT-FET). *IEEE Transactions on Electron Devices* 67(6): 2593–2599.

40. Jo, J. and Shin, C. (2016) Negative capacitance field effect transistor with hysteresis-free sub-60-mV/decade switching. *IEEE Electron Device Letters* 37(3): 245–248.

41. Jo, J. and Shin, C. Negative capacitance field effect transistor with hysteresis-free sub-60-mV/decade switching. *IEEE Electron Device Letters* 37(3): 245–248

42. Lee, M. H., et al. (2016) Physical thickness 1 x nm ferroelectric HfZrOx negative capacitance FETs. In: *2016 IEEE International Electron Devices Meeting (IEDM)*. IEEE.

43. Kobayashi, M., et al. (2017) Negative capacitance for boosting tunnel FET performance. *IEEE Transactions on Nanotechnology* 16(2): 253–258.

44. Seo, J., Lee, J. and Shin, M. Analysis of drain-induced barrier rising in short-channel negative-capacitance FETs and its applications. *IEEE Transactions on Electron Devices* 64(4): 1793–1798.

45. Wang, X., et al. (2019) Van der Waals negative capacitance transistors. *Nature Communications* 10(1): 1–8.

46. Zhou, J., et al. Negative differential resistance in negative capacitance FETs. *IEEE Electron Device Letters* 39(4): 622–625.

47. Yuan, Z. C., et al. (2016) Switching-speed limitations of ferroelectric negative-capacitance FETs. *IEEE Transactions on Electron Devices* 63(10): 4046–4052.

48. Kobayashi, M. (2018) A perspective on steep-subthreshold-slope negative-capacitance field-effect transistor. *Applied Physics Express* 11(11): 110101.

49. Khan, A. I., Keshavarzi, A. and Datta, S. The future of ferroelectric field-effect transistor technology. *Nature Electronics*. 3(10): 588–597.

50. Singh, K. J., et al. (2022) Understanding negative capacitance physical mechanism in organic ferroelectric capacitor. *Solid-State Electronics* 194: 1–5.

9 Mathematical Approach for the Foundation of Negative Capacitance Technology

Shiromani Balmukund Rahi[1], Abhishek Kumar Upadhyay[2], Young Suh Song[3], Nidhi Sahni[4], Ramakant Yadav[5], Umesh Chandra Bind[6], Guenifi Naima[7], Billel Smaani[8], Chandan Kumar Pandey[9], Samir Labiod[10], T.S. Arun Samul[11], Hanumant Lal[12], H. Bijo Josheph[13]

[1]Department of Electrical Engineering, Indian Institute of Technology, Kanpur 208016, India
[2]X-FAB Semiconductor foundries, 99097 Erfurt, Germany
[3]Korea Military Academy, Seoul, Republic of Korea
[4]Department of Mathematics, Sharda School of Basic Sciences and Research, Sharda University Greater Noida, Uttar Pradesh-201310, India
[5]Department of Electrical & Electronics Engineering, Mahindra University, Hyderabad 500043, India
[6]Centre of Nanotechnology, Indian Institute of Technology Roorkee, Roorkee 247667, India
[7]LEA Electronics Department, University Mostefa Benboulaid of Batna 2, Batna - 05000, Algeria
[8]Centre Universitaire Abdelhafid Boussouf – Mila, 43000, Algeria
[9]VIT-AP University, Amaravati, Andhra Pradesh India
[10]Laboratoire d'Automatique Appliquée, Université M'Hamed Bougara de Boumerdes, Algeria
[11]Department of ECE, National Engineering College, Kovilpatti, 628503, India
[12]Department of Electronics and Communication Engineering, Buddha Institute of Technology, Gorakhpur India
[13]Department of Physics and Nanotechnology, SRM Institute of Science and Technology Channi, India

DOI: 10.1201/9781003373391-9

9.1 GENERALIZED CLASSIFICATION OF FIELD EFFECT DEVICES BASED ON FERROELECTRIC

The continuous scaling with conventional MOSFET has reached a plateau. In respect of the continuous development with conventional scaling trends, FinFET is one of the best examples [1–5]. For further improvements in FinFET, experts have developed gate-all-around (GAA) FET structures with improved electrical advantages such as lowered short-channel FETs (SCEs), which however suffer from self-heating, such as SOI FETs. Self-heating has impaired the search for advancements in this technology [6–10].

Another steep subthreshold swing device category is negative capacitance FET, denoted as NCFET. This is a thin layer of ferroelectric material sandwiched with a conventional MOS structure like bulk NCFET [11–19]. In bulk NCFET, a thin layer of ferroelectric material is deposited with conventional dielectrics such as HFO_2 in a parallel combination. The ferroelectric material has negative capacitance and an applied electric field, and is discussed in the following section in detail [20–25]. Impact ionization MOS has lower SS but higher voltage operation. This shows that there is less interest in low-voltage applications. Another feature of negative capacitance with conventional MOS devices is that it acts as a voltage amplifier [26–29]. A thin layer helps the designer accomplish amplification of the vertical electric field (also known as the E-field) which the transistor perceives [30–34]. NCFET strategically incorporates a ferroelectric (FE) layer within the transistor's gate, which also amplifies the voltage, thereby causing NCFET to run at a lower voltage while sustaining its performance with remarkable energy savings.

Ferroelectric materials are a class of dielectric substances that possess a unique property called spontaneous electric polarization, which means they have a net electric dipole moment even in the absence of an external electric field. This polarization can be reversed and realigned when subjected to an external electric field, exhibiting hysteresis behavior. This remarkable characteristic makes ferroelectric materials valuable for various technological applications, including but not limited to non-volatile memories, sensors, actuators, and electro-optic devices. The development and understanding of ferroelectric materials have opened up new avenues in nanoelectronics and advanced material science, offering promising possibilities for future electronic devices and cutting-edge technologies.

The application and investigation of negative capacitance in other devices has been continuous. FET devices with negative capacitance include NC FinFET, NC tunnel FET, NC-GAA, and negative capacitance NS FET. Figure 9.1 presents some of the most popular semiconductor industry developments for VLSI circuit and system applications. The semiconductor industry has utilized the advantages of conventional MOSFET. In the series development of conventional MOSFET, FinFET and gate-all-around (GAA) has been popular inventions having better gate control, with less short-channel FETs (SCEs) having a subthreshold swing (SS) greater than 60 mV/decade. Conventional MOSFETs and related FET devices that have been developed show less suitability for future uses due to ultra-scaled applications. Nanosheet FETs like FinFET and GAA having better scalability and survivability for future have been suggested as possible solutions. Another category of MOS devices has SS lower than 60 mV/decade. Tunnel

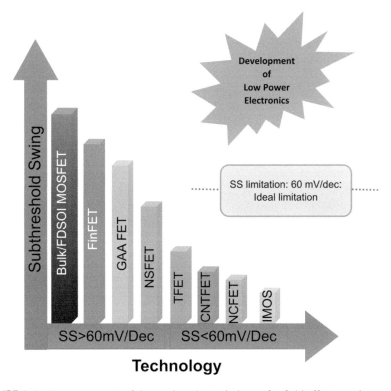

FIGURE 9.1   Progress report of the semiconductor industry for field effect transistors.

NCFET, NC-tunnel FET, and IMOS are common low subthreshold swing devices. A rough classification of these based on SS is shown in Figure 9.2 [20–35].

Figure 9.2 shows some advanced developments based on negative capacitance research in FinFET, GAA, and nanosheet FETs. The short form of negative capacitance, 'NC,' have been added to newly developed FETs known as NC FETs for conventional bulk MOSFET. Meanwhile the NC-tunnel FET is used for negative capacitance-based tunnel FETs. Here both m- and n-factors have been optimized by using band-to-band tunnel transport and ferroelectric material in the conventional gate region. Similarly, NC-FinFET, NC-GAA, and negative capacitance nanosheets are more advanced developments.

For TFET-based circuit and system applications, the device design element should have maximum I_{ON}, lowest possible SS, and suppressed ambipolar current I_{amb}. On-state switching current (I_{ON}) with possible steep subthreshold swing (SS) having minimal ambipolar current (I_{amb}) are some of the critical design goals for circuit designers for ultra-low-power applications.

$$T(E) \propto \left(-\frac{4\sqrt{2m^*}E_{g-effective}^{\frac{3}{2}}}{3|q|\hbar\left(E_{g-effective}+\Delta\Phi\right)}\sqrt{\frac{\varepsilon_{si}}{\varepsilon_{ox}}}t_{ox}t_{si} \right)\Delta\Phi, \text{ the tunneling mechanism in the}$$

device, in short $T(E)$, is used to represent the quantum band-to-band tunneling. This

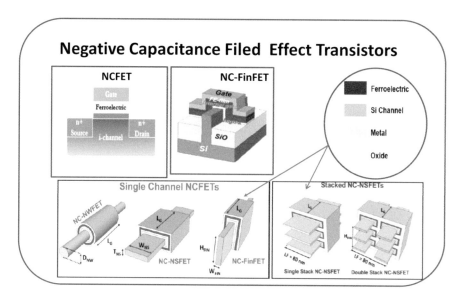

FIGURE 9.2   Progress in NCFET technology with advanced FETs.

has been developed with the help of Wentzel–Kramers–Brillouin (WKB) with triangular approximation, where, m^* is the effective mass, $E_{g\text{-effective}}$ is the effective band gap, $\Delta\Phi$ is the tunneling window, ε_{si} and ε_{ox} are the dielectric constants of Si and oxide material, respectively, and t_{si} and t_{ox} are the thickness of Si and oxide material, respectively.

9.2   MATHEMATICAL MODELING OF NEGATIVE CAPACITANCE SCIENCE AND TECHNOLOGY

In the field of electrical engineering, capacitance (C) has been traditionally modeled as in Equation (9.1). In Equation (9.1), the variable Q is used for charge that is stored on the capacitor's conductive plates and V is the applied voltage.

$$C = \frac{dQ}{dV}$$ (9.1)

For CMOS technology, capacitance formed between the gate region is modeled as shown in Equation (9.2).

$$C_{MOS} = \frac{dQ_t}{d\left(V_{IG} - V_{SUB}\right)}$$ (9.2)

As illustrated in Equation (9.2), the variable Q_t represents the total charge of the intermediate gate (technically, it is also the absolute value which is the same as the charge stored in channel layer Q_i) which is equal to the charge in the ferroelectric thin layer, used on top of the conventional gate dielectric material, that is, $Q_t = Q_{FE} = |Q_i|$. Similarly, the FE layer capacitance can be defined as follows:

$$C_{FE} = \frac{dQ_i}{dV_{FE}} \tag{9.3}$$

Equation (9.3) shows V_{FE}, which is the voltage drop in ferroelectric given by the Landau-Khalantikov (L-K) equation in the static state. Here, V_G is the gate terminal voltage. C_{fe} is the equivalent capacitance produced in the used ferroelectric thin layer. The variable C_{int} is used for the equivalent in base MOSFET. The variables *Cox*, *Cs*, and *ψs* have been used for oxide capacitance, surface potential, and body capacitance, respectively. It has been experimentally proved that the surface potential of negative capacitance FETs is amplified due to existence of negative capacitance in ferroelectric materials. While the idea of the NCFET could be seen as somewhat simple in theory, there has been consistent difficulty in finding and discovering a suitable ferroelectric material that can be integrated into advanced MOSFETs. Indeed, most known ferroelectric materials cannot be used, because they are not CMOS-compatible technologies. Figure 9.3 presents a sketch diagram for an equivalent capacitance model for conventional NCFETs.

As is known for conventional MOSFETs, there is a drain on the source current, $I_{DS} = \int(V_{GS}, V_{DS})$, but in the NCFET s, $I_{DS-NC} = \int(V_{GS}, V_{FE}, V_{DS})$. The voltage across ferroelectric material has been modeled in Equation (9.4). As shown in this equation, the voltage across FE material depends on the T_{FE} thinness and Landau parameters, α, β, and γ.

$$V_{FE} = \left(2\alpha Q_t + 4\beta Q_t + 6\gamma Q_t\right)T_{FE} \tag{9.4}$$

where the variables α, β, and γ are Landau coefficients of the ferroelectric materials and T_{FE} is the thickness. Using Equation (9.4), C_{FE} can be calculated with the help of Equation (9.5)

$$C_{FE} = \frac{1}{\left(2\alpha Q_t + 4\beta Q_t + 6\gamma Q_t\right)T_{FE}} \tag{9.5}$$

The potential barriers of the polarization switching pathways are also among the most important properties of ferroelectric materials, which are determined using the nudged elastic band method (NEB). The polarization of bulk systems was calculated using the Barry phase approach. The Landau free energy (U), estimated by Equation (9.6), of the ferroelectric materials is defined using the Landau-Khalantikov theory.

$$U = \alpha P^2 + \beta P^4 + \gamma P^6 - E.P \tag{9.6}$$

where the variables α, β, and γ are the Landau coefficients of the ferroelectric materials, which vary with biaxial strain. In Equation (9.6), E is the external field.

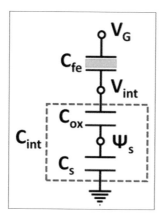

FIGURE 9.3 Capacitance equivalent model for negative capacitance FETs.

The L-K variables in Equation (9.6) vary with biaxial strain. When a material is under the ferroelectric phase, it starts making a stable polarization state which minimizes its free energy (U) landscape. These stable states, that is $\left(\frac{\partial^2 U}{\partial D^2}\right) > 0$, indicate positivity permittivity $\left(\in \text{ is proportional to } \left(\frac{\partial^2 U}{\partial D^2}\right)^{-1}\right)$, where D is the electric displacement field as shown in Figure 9.4. Pr is the remnant polarization, E is the applied electric field, and E_c is the critical electric field.

Figure 9.5 presents the capacitance model for negative capacitance FET devices. As shown in Figure 9.4, a thin layer of ferroelectric materials is sandwich on the top of conventional dielectric materials like HfO_2 [40–59]. The key factor to understand the overall principle of the negative capacitance effect is that the negative differential capacitance of ferroelectric ($C_{EF}<0$) compensates for the positive capacitance ($C_{MOS}>0$) in the device such that the resulting gate capacitance, $C_G = \left(C_{FE}^{-1} + C_{MOS}^{-1}\right)^{-1}$, can be made larger than C_{MOS} [60–62].

The ferroelectric layer, which amplifies the gate voltage of the baseline MOSFET, was modeled based on the L-K equation, as defined in Equation (9.7).

$$\rho\frac{dP}{dt} + \nabla_p U = 0 \qquad (9.7)$$

where the variables ρ and P are the resistivity and polarization, respectively. When the ferroelectric material is under a steady state, $\left(\frac{dP}{dt} = 0\right)$, E can be written as shown in Equation (9.8)

$$E = 2\alpha P + 4\beta P^3 + 6\gamma P^5 \qquad (9.8)$$

which is induced through polarization of the ferroelectric layer. Negative capacitance is available in a series connection of a ferroelectric and normal capacitor

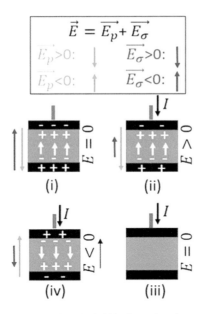

FIGURE 9.4   Impact of an applied electric field in ferroelectric materials.

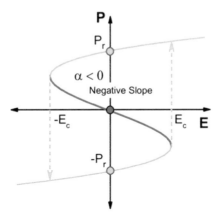

FIGURE 9.5   Verification of negative capacitance in ferroelectric materials.

$$\frac{1}{C_{Eq}} = \frac{1}{C_{FE}} + \frac{1}{C_{Base-MOS}}$$
(9.9)

In Equation (9.9), C_{Eq} is the equivalent capacitance, C_{FE} is the capacitance due to ferroelectric material, and $C_{Base-MOS}$ is the base capacitance/conventional capacitance available in normal MOSFET devices, like bulk MOSFET.

One of the most important aspects of NCFETs is that they have a large internal amplification factor, and the β, is known as the lowering of the body factor, because

$C_{ox}\langle 0 ; |C_{ox}|\rangle C_s$ in NCFETs and the body factor 'm' followed by Equation (9.10), which directly influence the SS

$$m = \frac{1}{\beta} \times \frac{\partial V_{int}}{\partial \varphi_s} = \frac{1}{\beta} \times \left(1 + \frac{C_s}{C_{ox}}\right) \ll 1 \qquad (9.10)$$

For NCFET technology, using the amplification factor, $\beta > 1$, the SS of an NCFET is defined by the following equation:

$$SS_{nc} = \left(\frac{\partial \log I_{DS}}{\partial V_{GS}}\right)^{-1} = \frac{\partial V_{int}}{\partial \log I_{DS}} \times \frac{\partial V_{GS}}{\partial V_{int}} = \frac{SS_{Bulk}}{\beta} \qquad (9.11)$$

9.3 CONCLUSION

In this chapter, various mathematical approaches have been introduced in order to advance the understanding of NCFET. Thanks to the physical mechanism of selected oxide layers, the NCFET could serve as the next-generation transistor, to be aimed for high-performance CPUs with sub-7 nm technology nodes. These ferroelectric materials could enable a huge breakthrough in modern semiconductor technology, where the limitation of scaling has put severe limitations on semiconductor development.

ACKNOWLEDGMENT

This chapter has been written by the corresponding author (Dr. Shiromani Balmukund Rahi). The author has written and edited this chapter by referring to various earlier research. The author greatly appreciates the relevant earlier research which helped the author in writing this manuscript.

REFERENCES

1. Iwai H. (2008) CMOS technology after reaching the scale limit. In: *Extended Abstracts – 2008 8th International Workshop on Junction Technology (IWJT '08)*, pp. 1–2.
2. Lundstrom, M. (2003) Device physics at the scaling limit: what matters? [MOSFETs]. In: *IEEE International Electron Devices Meeting 2003*, pp. 33.1.1–33.1.4.
3. Anas, M., Amin, S.I., Beg, M.T. et al. (2022) Design and analysis of GaSb/Si based negative capacitance TFET at the device and circuit level. *Silicon* 14: 11951–11961.
4. Roy, K., Mukhopadhyay, S. and Mahmoodi-Meimand, H. (2003) Leakage current mechanisms and leakage reduction techniques in deep-submicrometric CMOS circuits. *Proceedings of the IEEE* 91(2): 305–327.

5. Hiramoto, T. et al. (2016) Ultra-low power and ultra-low voltage devices and circuits for IoT applications. In: *2016 IEEE Silicon Nanoelectronics Workshop (SNW)*, pp. 146–147.
6. Makimoto, T. and Sakai, Y. (2003) Evolution of low power electronics and its future applications. In: *Proceedings of the 2003 International Symposium on Low Power Electronics and Design, 2003. ISLPED '03*, pp. 2–5.
7. Taur, Y. and Ning, T. H. (1998) *Fundamentals of Modern VLSI Devices*. New York: Cambridge University Press, ch. 2, pp. 94–95.
8. Roy, K., Mukhopadhyay, S. and Mahmoodi-Meimand, H. (2003) Leakage current mechanisms and leakage reduction techniques in deep-submicrometric CMOS circuits. *Proceedings of the IEEE* 91(2): 305–327.
9. Pierret, R. (1996) *Semiconductor Device Fundamentals*. Reading, MA: Addison-Wesley, ch. 6, pp. 235–300.
10. Taur, Y. and Ning, T. H. (1998) *Fundamentals of Modern VLSI Devices*. New York: Cambridge University Press, ch. 2, pp. 94–95.
11. Rabaey, J. M. (1996) *Digital Integrated Circuits*. Englewood Cliffs, NJ: Prentice-Hall, ch. 2, pp. 55–56.
12. Taur, Y. and Ning, T. H. (1998) *Fundamentals of Modern VLSI Devices*. New York: Cambridge University Press, ch. 3, pp. 143–144.
13. Paul, B. C., Soeleman, H. and Roy, K. (2001) An 8 × 8 subthreshold digital CMOS carry save array multiplier. In: *Proceedings of the 27th European Solid-State Circuits Conference (ESSCIRC '01)*, pp. 377–380, Villach, Austria, September 2001.
14. Singh, K. J., Bulusu, A. and Dasgupta, S. (2022) Origin of negative capacitance transient in ultrascaled multidomain metal-ferroelectric-metal stack and hysteresis-free Landau transistor. *IEEE Transactions on Electron Devices* 69(3): 1284–1292.
15. Wei, C. and Banerjee, K. (2020) Is negative capacitance FET a steep-slope logic switch? *Nature Communications* 11(1): 1–8.
16. Saeidi, A. et al. (2017) Negative capacitance as performance booster for tunnel FETs and MOSFETs: an experimental study. *IEEE Electron Device Letters* 38(10): 1485–1488.
17. Kobayashi, M., Jang, K., Ueyama, N. and Hiramoto, T. (2016) Negative capacitance as a performance booster for tunnel FET. In: *2016 IEEE Silicon Nanoelectronics Workshop (SNW)*, pp. 150–151.
18. Kim H. W. and Kwon, D. (2021) Gate-normal negative capacitance tunnel field-effect transistor (TFET) with channel doping engineering. *IEEE Transactions on Nanotechnology* 20: 278–281.
19. Lin, C.-I. et al. (2016) Effects of the variation of ferroelectric properties on negative capacitance FET characteristics. *IEEE Transactions on Electron Devices* 63(5): 2197–2199.
20. Cao, W. and Banerjee, K. (2020) Is negative capacitance FET a steep-slope logic switch? *Nature Communications* 11(1): 1–8.
21. Salahuddin, S. and Datta, S. (2008) Use of negative capacitance to provide voltage amplification for low power nanoscale devices. *Nano Letters* 8(28): 405–410.
22. Zhirnov, V. V. and Cavin, R. K. (2008) Negative capacitance to the rescue? *Nature Nanotechnology* 3(2): 77–78.
23. Khan, A. I. et al. Ferroelectric negative capacitance MOSFET: Capacitance tuning & antiferroelectric operation. In: *2011 International Electron Devices Meeting*. IEEE, 2011.
24. Catalan, G., Jiménez, D. and Gruverman, A. (2015) Negative capacitance detected. *Nature Materials* 14(2): 137–139.

25. Rahi, S. B., Tayal, S. and Upadhyay, A. K. (2021) A review on emerging negative capacitance field effect transistor for low power electronics. *Microelectronics Journal* 116: 105242.

26. Upadhyay, A. K., Rahi, S. B., Tayal, S. and Song, Y. S. (2022) Recent progress on negative capacitance tunnel FET for low-power applications: Device perspective. *Microelectronics Journal* 129: 105583.

27. Amrouch, H., van Santen, V. M., Pahwa, G., Chauhan, Y. and Henkel, J. NCFET to rescue technology scaling: Opportunities and challenges. In: *2020 25th Asia and South Pacific Design Automation Conference (ASP-DAC)*, pp. 637–644.

28. You, W.-X-, Su, P. and Hu, C. (2018) Evaluation of NC-FinFET based subsystem-level logic circuits using SPICE simulation. In: *2018 IEEE SOI-3D-Subthreshold Microelectronics Technology Unified Conference (S3S)*, pp. 1–2.

29. Sakib, F. I., Hasan, M. A. and Hossain, M. (2020) Exploration of negative capacitance in gate-all-around Si nanosheet transistors. *IEEE Transactions on Electron Devices* 67(11): 5236–5242.

30. Kim, M., Seo, J. and Shin, M. (2018) Biaxial strain based performance modulation of negative-capacitance FETs. In: *2018 International Conference on Simulation of Semiconductor Processes and Devices (SISPAD)*, pp. 318–322.

31. Hoffmann, M., Slesazeck, S. and Mikolajick, T. Progress and future prospects of negative capacitance electronics: A materials perspective. *APL Materials* 9(2): 020902.

32. Dong, Z. and Guo, J. (2017) A simple model of negative capacitance FET with electrostatic short channel effects. *IEEE Transactions on Electron Devices* 64(7): 2927–2934.

33. Li, X., et al. Enabling energy-efficient nonvolatile computing with negative capacitance FET. *IEEE Transactions on Electron Devices* 64(8): 3452–3458

34. Kwon, D., et al. (2019) Negative capacitance FET with 1.8-nm-thick Zr-doped HfO_2 oxide. *IEEE Electron Device Letters* 40(6): 993–996.

35. Li, K.-S., et al. (2015) Sub-60mV-swing negative-capacitance FinFET without hysteresis. In: *2015 IEEE International Electron Devices Meeting (IEDM)*. IEEE.

36. Kobayashi, M., et al. Experimental study on polarization-limited operation speed of negative capacitance FET with ferroelectric HfO_2. In: *2016 IEEE International Electron Devices Meeting (IEDM)*. IEEE.

37. Tayal, S., Rahi, S. B., Srivastava, J. P. and Bhattacharya, S. (2021) Recent trends in compact modeling of negative capacitance field-effect transistors. In: *Semiconductor Devices and Technologies for Future Ultra Low Power Electronics*, pp. 203–226. CRC Press.

38. Tayal, S., Upadhyay, A. K., Kumar, D. and Rahi, S. B. (eds.) (2022) *Emerging Low-Power Semiconductor Devices: Applications for Future Technology Nodes*. CRC Press.

39. Hu, V. P.-H., et al. (2020) Optimization of negative-capacitance vertical-tunnel FET (NCVT-FET). *IEEE Transactions on Electron Devices* 67(6): 2593–2599.

40. Jo, J. and Shin, C. (2016) Negative capacitance field effect transistor with hysteresis-free sub-60-mV/decade switching. *IEEE Electron Device Letters* 37(3): 245–248.

41. Jo, J. and Shin, C. (2016) Negative capacitance field effect transistor with hysteresis-free sub-60-mV/decade switching. *IEEE Electron Device Letters* 37(3): 245–248.

42. Lee, M. H., et al. (2016) Physical thickness x nm ferroelectric HfZrOx negative capacitance FETs. In: *2016 IEEE International Electron Devices Meeting (IEDM)*. IEEE.

43. Kobayashi, M., et al. (2017) Negative capacitance for boosting tunnel FET performance. *IEEE Transactions on Nanotechnology* 16(2): 253–258.

44. Seo, J., Lee, J. and Shin, M. (2017) Analysis of drain-induced barrier rising in short-channel negative-capacitance FETs and its applications. *IEEE Transactions on Electron Devices* 64(4): 1793–1798.

45. Wang, X., et al. (2019) Van der Waals negative capacitance transistors. *Nature Communications* 10(1): 1–8.

46. Zhou, J., et al. (2018) Negative differential resistance in negative capacitance FETs. *IEEE Electron Device Letters* 39(4): 622–625.

47. Yuan, Z. C., et al. (2016) Switching-speed limitations of ferroelectric negative-capacitance FETs. *IEEE Transactions on Electron Devices* 63(10): 4046–4052.

48. Kobayashi, M. (2018) A perspective on steep-subthreshold-slope negative-capacitance field-effect transistor. *Applied Physics Express* 11(11): 110101.

49. Khan, A. I., Keshavarzi, A. and Datta, S. (2020) The future of ferroelectric field-effect transistor technology. *Nature Electronics* 3(10): 588–597.

50. Singh, K. J. et al. (2022) Understanding negative capacitance physical mechanism in organic ferroelectric capacitor. *Solid-State Electronics* 194: 108350.

51. Khan, A. I., et al. (2015) Negative capacitance in a ferroelectric capacitor. *Nature Materials* 14(2): 182–186.

52. Singh, K. J., Bulusu, A. and Dasgupta, S. Multidomain negative capacitance effect in P(VDF-TrFE) ferroelectric capacitor and passive voltage amplification. *IEEE Transactions on Electron Devices* 67(11): 4696–4700.

53. Obradovic, B., Rakshit, T., Hatcher, R., Kittl, J. and Rodder, M. S. (2018) Modeling of negative capacitance of ferroelectric capacitors as a nonquasi static effect. *Mesoscale and Nanoscale Physics* 13: 1–8.

54. Singh, K. J., Bulusu, A. and Dasgupta, S. (2022) Origin of negative capacitance transient in ultrascaled multidomain metal-ferroelectric-metal stack and hysteresis-free landau transistor *IEEE Transactions on Electron Devices* 69(3): 1284–1292.

55. Rusu, A. and Ionescu, A. M. (2012) Analytical model for predicting subthreshold slope improvement versus negative swing of S-shape polarization in a ferroelectric FET. In: *Proceedings of MIXDES 2012*, Warsaw, pp. 55–59.

56. Singh, K. J., Chauhan, N., Bulusu, A. and Dasgupta, S. (2022) Physical cause and impact of negative capacitance effect in ferroelectric P(VDF-TrFE) gate stack and its application to landau transistor. *IEEE Open Journal of Ultrasonics, Ferroelectrics, and Frequency Control*, 2: 55–64.

57. Singh, K. J., Bulusu, A. and Dasgupta, S. Harnessing maximum negative capacitance signature voltage window in P(VDF-TrFE) gate stack. In: *Proceedings of the IEEE International Symposium on Circuits and Systems (ISCAS)*, May 2021, pp. 1–5.

58. Shin, C. (2019) Experimental understanding of polarization switching in PZT ferroelectric capacitor. *Semiconductor Science and Technology* 34(7): 075004.

59. Singh, K. J., Bulusu, A. and Dasgupta, S. (2021) Ultrascaled multidomain P(VDF-TrFE) organic ferroelectric gate stack to the rescue. In: *Proceedings of the IEEE Latin America Electron Devices Conference (LAEDC)*, April 2021, pp. 1–4.

60. Li, Y., Lian, Y. and Samudra, G. S. (2015) Quantitative analysis and prediction of experimental observations on quasi-static hysteretic metal–ferroelectric–metal–insulator–semiconductor FET and its dynamic behaviour based on Landau theory. *Semiconductor Science and Technology* 30(4) 045011.

61. Müller, J., et al. (2011) Ferroelectric Zr 0.5 Hf 0.5 O_2 thin films for nonvolatile memory applications. *Applied Physics Letters* 99(11): 112901.

62. Alam, M.N.K., Roussel, P., Heyns, M. et al. (2019) Positive non-linear capacitance: the origin of the steep subthreshold-slope in ferroelectric FETs. *Scientific Reports* 9: 14957.

Index